THE LOST TECHNOLOGIES OF THE ANCIENT WORLD

Kayla Brooks

The Lost Technologies of the Ancient World

Kayla Brooks

Published by Kayla Brooks, 2026.

First published 2025

Printed in the United States of America

While every precaution has been taken in the preparation of this book, the publisher assumes no responsibility for errors or omissions, or for damages resulting from the use of the information contained herein.

THE LOST TECHNOLOGIES OF THE ANCIENT WORLD

First edition. March 12, 2026.

ISBN: 979-8233595936

Written by Kayla Brooks.

Table of Contents

For the builders whose names were never written down.

Contents

Introduction
The Genius We Forgot

In the autumn of 1900, a team of Greek sponge divers working the waters between the island of Antikythera and the Peloponnesian coast made a discovery that would eventually overturn nearly everything the modern world believed about the technological sophistication of the ancient Greeks. Their boat had been blown off course, and while waiting for the weather to turn, the divers descended into the deep water near the island and found the wreck of a Roman cargo ship that had gone down sometime around 60 BCE. The vessel was loaded with treasures: bronze and marble statues, glassware, jewellery, amphorae. Over the following months, Greek authorities worked to salvage what they could, and among the corroded, encrusted fragments pulled from the seabed was a lump of bronze about the size of a large dictionary, fractured and shapeless with corrosion, that at first attracted little notice.

When researchers examined it more carefully over the following years, they began to understand that what they were holding was not mere decoration but a machine. Not a gear in isolation, which might have suggested some routine mechanical fitting, but gears: interlocking, precisely cut bronze gears of different sizes, meshed together in a configuration that implied systematic engineering intelligence of the kind that produces calculating instruments rather than simple tools. The Antikythera Mechanism, as it came to be known, was not supposed to exist. According to the received understanding of ancient technology, nothing like it should have been possible until the Renaissance at the earliest, and possibly not until the industrial era. And yet here, preserved in saltwater for more than two thousand years, was proof that someone in the ancient world had possessed mathematical knowledge, mechanical skill, and engineering precision that the modern era had completely forgotten they ever had.

That mechanism is where this book begins, but it is far from where this book ends. The Antikythera device is the most famous example of a much wider and more unsettling phenomenon: the ancient world was, in critical respects, far more technically advanced than it has been given credit for, and a great deal of what it knew, built, and discovered is gone. Not misplaced. Not patiently archived somewhere awaiting recovery. Actually lost, in the deepest sense: vanished through fire, war, economic collapse, the death of workshops and craft traditions, the erasure of languages, the abandonment of cities, and the slow grinding entropy of civilizations that rise and fall and leave only fragments of themselves behind.

This book is an investigation into those fragments. It is an attempt to understand not only what the ancient world built and discovered, but how, why those achievements have so often been misunderstood or underestimated, what has been recovered through modern archaeology and materials science, and what remains genuinely debated or irretrievably gone. It does not belong to the literature of conspiracy or pseudoscience. It does not traffic in claims about alien architects, lost super civilizations, or secret energies. The real story is far more interesting than any of those fictions, and it asks considerably more of the reader in return. Understanding genuine ancient technical achievement requires patience, a willingness to sit with complexity, and an ability to distinguish between what is known, what is plausible, and what is merely appealing.

The chapters that follow examine twelve different technologies or technical traditions from the ancient world, each one chosen because it reveals something distinctive about the relationship between human ingenuity and historical memory. Some of these achievements, like the Antikythera Mechanism or Roman marine concrete, are now well documented through meticulous scholarship and laboratory analysis, and the wonder they provoke is grounded in solid evidence. Others, like the composition of Greek

Fire or the precise methods used to cut and transport the largest Egyptian stones, remain partially contested, surrounded by serious scholarly debate that the historical record only partially resolves. And a few, like the so-called Baghdad Battery, sit in genuinely uncertain territory, where careful critical thinking is more valuable than any confident claim in either direction.

What unites all twelve is that they challenge a deeply embedded assumption that the modern world tends to carry about the past: the assumption that progress moves in a single direction, that earlier automatically means simpler, and that civilizations which lacked our technology must therefore have lacked our sophistication. This assumption is almost always wrong, and it is wrong in ways that say as much about the modern world's self-image as they do about the ancient world's capabilities.

Consider what the Roman engineers of the first and second centuries CE actually understood about materials science, even without the vocabulary or theoretical framework we now use to describe it. They knew from generations of empirical observation and practical experiment that mixing volcanic ash from the region around Pozzuoli with seawater and lime produced a concrete that, rather than degrading when exposed to marine environments, actually grew stronger over time as a result of a chemical reaction between the seawater and the mineral crystals forming within the mix. They did not call it that, and they had no chemical theory available to explain why it worked. But they understood the effect. They systematically exploited it. They built harbour structures and seawalls using this material that have outlasted virtually everything the modern world has built with Portland cement in equivalent conditions. Roman concrete from harbours along the Italian coast has been sitting in the Mediterranean for two thousand years and is in some respects still getting stronger. Modern concrete seawalls typically show significant degradation within fifty years. That is not mythology. That is materials science.

Or consider the blade we call Damascus steel, the legendary material that medieval Islamic smiths produced from Indian wootz steel imported along overland and maritime trade routes from the subcontinent. Blades made from this material held an edge of extraordinary sharpness, displayed a visible surface pattern of flowing crystalline structures, and demonstrated a combination of hardness and flexibility that would be enviable in modern metallurgy. Centuries of European crusaders and travellers wrote about these weapons with something approaching awe. When the craft tradition that produced them was disrupted in the early eighteenth century, partly through the exhaustion of particular ore sources from India, the exact method by which the distinctive microstructure was achieved was lost. Modern metallurgists have worked for decades to understand the chemistry involved, and while they have made significant progress, the precise combination of skill, timing, temperature control, and material sourcing that produced the finest historical examples remains only partially reconstructed.

A blade made of wootz steel involved knowledge that lived in a smith's hands and eyes and judgment as much as it lived in any describable recipe. The master stood at his forge for hours, watching the colour of the hot metal shift from black through cherry red to the bright straw yellow that told him the temperature was exactly right, feeling in the rhythm of his hammer blows the resistance of a material he had spent years learning to read. That kind of embodied knowledge is among the most fragile things in human culture. It cannot be fully captured in any text, and when the workshops that carried it closed, it went with them.

These are not isolated curiosities. They are symptoms of a pattern in the history of human knowledge that deserves sustained, serious attention. Brilliant technical achievements have been lost before, and they will be lost again. The conditions that produce their loss are not exceptional or mysterious; they are the ordinary conditions of history, the disruption of trade routes that supplied essential raw materials, the collapse of patronage systems that supported specialist craftsmen, the destruction of archives that preserved technical instructions, the decline of languages in which those instructions were written, the conquest of cities whose workshops contained expertise that existed nowhere else.

The Library of Alexandria is the most famous example in the ancient world, though its history is considerably more complex than the popular image of a single catastrophic burning. The library was damaged, partially recovered, damaged again, absorbed into shifting political structures, and ultimately declined as the city and the patronage networks that sustained it changed over several centuries of late antiquity. What was lost when it finally ceased to function as a living intellectual centre was not just texts but the community of scholars who read, debated, annotated, and built upon those texts, the network of institutional memory that gave the documents their living context. You can sometimes reconstruct a recipe from a surviving manuscript, but you cannot reconstruct the workshop that made sense of it.

This points toward something that any honest account of ancient technology must grapple with seriously: the difference between written knowledge and embodied knowledge, between what can be put into a text and what lives in the hands and trained judgment of a practitioner. A medieval glassmaker did not need to write down the temperature at which a particular batch of glass became workable, or the way a good batch sounded when you tapped it at the right moment, or the subtle colour shift in the furnace that told an experienced eye the conditions were exactly right. She knew those things the way a skilled cook knows when bread is ready without consulting a timer. That knowledge was transmitted through apprenticeship, through years of proximity to a master, through the slow accumulation of experience that no text can fully capture. When the workshop closed, when the master died without training a successor, when the guild that organized the trade was disrupted by war or pestilence or economic collapse, that knowledge did not go into storage waiting to be retrieved. It died.

Modern archaeology has made extraordinary progress in recovering at least a partial understanding of what was lost. Experimental archaeologists have rebuilt working reconstructions of Roman catapults and water organs and Greek musical instruments. Materials scientists have analyzed the microstructure of ancient metals and the chemical composition of ancient pigments. Engineers have studied the tolerances achieved in Egyptian stonework and found that some surfaces were fitted to a fraction of a millimetre over spans of many metres. Digital reconstruction teams have used computed tomography scanning to look inside the corroded fragments of the Antikythera Mechanism without damaging them, revealing gear teeth and inscriptions invisible for two thousand years. The picture that emerges is of ancient civilizations that were not primitive, not approximate, and not merely lucky. They were systematic, experimental, observant, and in many domains extraordinarily skilled.

What they often lacked was not intelligence or technical capability but rather the particular combination of industrial infrastructure and abstract scientific theory that defines modern technology. Ancient engineers could not build a jet engine, not because they were less intelligent than modern engineers, but because they lacked access to the centuries of accumulated theoretical and industrial development that makes jet engines possible. What they could do, within their own material and intellectual circumstances, was frequently astonishing, and the fact that they did it without our tools, our energy sources, or our theoretical frameworks makes it more impressive rather than less. To explain Roman concrete by appealing to alien intervention or secret supernatural knowledge is to insult the Roman engineers. What they achieved, they achieved through centuries of patient empirical observation, systematic experimentation, and the accumulation of practical wisdom across generations of practitioners who pushed each other to do better.

This book is organized to take the reader through twelve distinct windows onto ancient technical achievement, each one illuminating a different facet of what was accomplished and what was lost. The early chapters focus on the objects and materials that most dramatically challenge modern assumptions: the Antikythera Mechanism, Roman concrete, Damascus steel, Greek Fire, and the deeply contested Baghdad Battery. The middle chapters broaden the lens to look at entire engineering traditions: Egyptian stoneworking, the Nazca lines, the automata of Hero of Alexandria, the hydraulic systems of the ancient urban world. The later chapters consider the grand problem of megalithic and polygonal masonry, then turn to the question of how knowledge was stored, transmitted, and lost, and what the ongoing revolution in archaeology continues to uncover. A final chapter sits with the open questions, the things that genuinely remain unknown or debated, and thinks about what that uncertainty means for how we understand the past.

Throughout all of it, the book tries to maintain a distinction that matters enormously and is often abandoned in popular writing on this subject: the distinction between what is well established, what is plausible but debated, and what is speculative or unsupported. Ancient technology attracts more than its share of sensational claims, and the sensationalism almost invariably obscures rather than illuminates the genuine wonder of what was actually accomplished. The real achievements do not need exaggeration. They need careful explanation, honest assessment of the evidence, and the kind of contextual thinking that places each achievement inside the civilization that produced it.

There is also a question lurking behind all of this that the book returns to at intervals: what does the loss of ancient technical knowledge tell us about the fragility of our own? The modern world has produced extraordinary technology, some of it of a complexity that would have been literally inconceivable to even the most brilliant ancient engineer. But modern technology is also, in certain respects, more vulnerable than ancient technology, more dependent on continuous infrastructure, more specialized, more interdependent, more exposed to disruption at any point in a long chain. The ancient world's most durable knowledge was embedded in physical craft traditions, in the hands of thousands of practitioners distributed across many workshops in many cities. Modern knowledge is in many ways more centralized, more abstract, more dependent on specialized equipment to access. That is not an argument for despair, and it is not a case for romanticizing the past. It is an observation worth holding in mind as each chapter unfolds.

Ancient life was hard, often brutal, frequently short, and largely without the medical, nutritional, and material comforts that modern technology provides. The point is not that the ancient world was better but that it was more sophisticated than it is usually given credit for, and that the mechanisms by which its sophistication was lost are not as remote from the present as modern confidence tends to assume.

The chapters ahead are full of remarkable things. There are gears no larger than a fingernail that tracked the cycles of eclipses with astronomical precision. There are concrete structures built two thousand years ago that are still hardening in the sea. There are sword blades whose microstructure contains features that materials scientists using electron microscopes did not understand until the twenty-first century. There are stone surfaces fitted together with tolerances that challenge experienced modern stonemasons. There are hydraulic systems in ancient cities that delivered water to thousands of households at pressure, raised from rivers and aquifers by mechanisms whose engineering is still studied by historians of technology. There are enormous lines drawn on a Peruvian plateau that required sophisticated surveying mathematics and organized labour on a scale implying administrative systems of considerable complexity, all without a single piece of modern equipment to guide the work. And behind all of these achievements, there are human beings: observant, patient, experimental, competitive, proud of their craft, and in possession of knowledge that no one alive today fully understands.

That is the story this book tells. Not the story of magical ancients with powers beyond our understanding, but the story of real people doing extraordinary things with the materials, methods, and accumulated knowledge of their time, and of all that their heirs, through accident and conflict and the ordinary disasters of history, allowed to slip away. What follows is an invitation to take those people seriously, to understand what they actually accomplished, and to sit with the genuine unease that comes from recognizing how much of what they knew is simply and permanently gone.

One

The Antikythera Mechanism and the First Computer

The ship went down in bad weather, and whatever its cargo was worth in the markets of Rome, the sea took all of it. For roughly two thousand years the wreck lay undisturbed on the sandy floor of the Aegean near the small island of Antikythera, in the strait that separates Crete from the Greek mainland, in water deep enough to keep it invisible and inaccessible to anyone without a diving suit. When a crew of Greek sponge divers stumbled upon it in October 1900, sheltering from a storm and using the unexpected delay to work a promising patch of seafloor, they found the wreck of a Roman cargo ship that had gone down sometime around 60 BCE. The vessel was loaded with treasures: bronze and marble statues, glassware, jewellery, amphorae. Among the corroded, encrusted fragments pulled from the seabed over the following months was a lump of bronze about the size of a large dictionary, fractured, encrusted with salt crystals and marine growth, and so visually unimpressive that it was set aside without particular ceremony while the more obviously beautiful objects received attention.

That lump sat in the National Archaeological Museum in Athens for more than a year before anyone looked at it carefully enough to notice that it contained gears. Not a gear, singular, which might have suggested a simple hinge or mechanical fitting of some routine kind. Gears. Plural. Interlocking, precisely cut bronze gears of different sizes, meshed together in a configuration that implied not just mechanical function but mechanical complexity, systematic design, and the kind of engineering intelligence that produces machines rather than mere tools. The historian and philologist Valerios Stais noticed the first gear teeth in 1902, and even then, for several more decades, no one could quite agree on what the object was or how it worked. It was too degraded, too fragmented, too encrusted with two millennia of marine chemistry to be immediately legible. Scholars proposed a variety of functions: an astrolabe of unusual design, a navigational instrument, a mechanical display piece. The debate dragged on inconclusively for decades.

It was the British science historian Derek de Solla Price who, in the 1950s and 1960s, produced the first rigorous analysis of the mechanism's internal structure and advanced a clear argument for what it actually was: an analogue computing device, built to model and predict the movements of celestial bodies with a precision that required not just mechanical ingenuity but deep knowledge of astronomy and mathematics. Price worked with X-ray photographs of the surviving fragments, three major pieces and dozens of smaller ones, and painstakingly reconstructed what he could of the gear train. His 1974 monograph, Gears from the Greeks, established the basic framework for understanding the device that all subsequent researchers have built upon, argued with, refined, and extended. It also contained a sentence that captures the cultural shock the mechanism continues to deliver: Price wrote that his reconstruction suggested a level of technological sophistication that had simply not been expected from the ancient world and that required a fundamental revision of assumptions about what Greek mechanical science had actually achieved. He was not overstating it.

The revision has only deepened with time. In 2005 and 2006, the Antikythera Mechanism Research Project, an international collaboration involving scholars from Greece, Britain, and the United States working alongside imaging specialists from Hewlett-Packard and the custom X-ray tomography firm X-Tek Systems, subjected the surviving fragments to a new generation of imaging technology. Polynomial texture mapping allowed the surface inscriptions, worn almost to invisibility, to be read as never before. High-resolution computed tomography scanning produced three-dimensional images of the internal gear structure with sufficient detail to count individual teeth and measure the angles of gear faces with a precision that Price, working from flat X-ray photographs, could never have approached.

The device, as it is now understood, was housed in a wooden box roughly the size of a shoebox, with bronze dials on its front and back faces and a system of at least thirty and possibly as many as thirty-seven interlocking bronze gears inside. The front dial displayed the positions of the sun and moon in the zodiac, along with a subsidiary dial showing the Egyptian calendar year. The two dials on the back face were more astronomically complex. The upper back dial tracked the Metonic cycle, a period of nineteen years after which the phases of the moon return to the same dates in the solar calendar, a cycle discovered by the Athenian astronomer Meton in the fifth century BCE and of great practical importance for synchronizing lunar and solar calendars. Nested within the Metonic dial was a subsidiary dial tracking the Callippic cycle, a seventy-six-year period that refines the Metonic calculation.

Below this, the lower back dial tracked the Saros cycle, a period of approximately eighteen years and eleven days after which lunar and solar eclipses recur in the same sequence, and a subsidiary dial tracked the Exeligmos cycle, the triple Saros of fifty-four years that brings eclipses back to roughly the same geographical position. Each of these astronomical periods required its own gear ratio, its own precisely sized wheel within the train, its own relationship to every other moving part. To hold all of this inside a portable wooden box was an act of miniaturization as much as it was an act of astronomical knowledge.

The precision required to encode these cycles in bronze gear ratios is genuinely extraordinary. The Metonic cycle requires a gear ratio of 254 to 19, meaning that one gear must turn 254 times for every 19 turns of another, because the moon completes 254 sidereal months in the same period that the sun completes 19 tropical years.

Achieving that ratio mechanically, with gears small enough to fit inside a portable box, requires not just mathematical knowledge of the cycle itself but the engineering skill to cut gear teeth with sufficient accuracy that the accumulated error across dozens of meshing pairs does not corrupt the output over years of use. The gear that drives the Metonic calculation in the Antikythera Mechanism has 223 teeth. Some of the smaller gears in the train have as few as 15 or 19. The tooth profiles, as revealed by the tomographic imaging, are cut with a consistency and accuracy that astonished the mechanical engineers who examined them, not perfect by modern standards, but far closer to precise than the supposed technological limitations of ancient Greek metalworking had led anyone to expect.

There is also an epicyclic gear system within the mechanism, a configuration in which one gear rotates around the outside of another while itself rotating on its own axis, a mechanical arrangement that allows the device to model the anomalous motion of the moon as understood within ancient astronomical theory. The Greeks knew from long observation that the moon does not move at a constant speed through the sky; it accelerates and decelerates in a regular pattern that the astronomer Hipparchus had described in the second century BCE using an epicyclic model. The Antikythera Mechanism encodes that model mechanically, using a small pin on one gear engaging a slot in an adjacent plate to produce a variable-speed output that mirrors the observed irregularity of lunar motion. Someone designed this feature deliberately, which means they understood Hipparchan astronomy well enough to translate its mathematical structure into bronze.

The inscriptions recovered by the polynomial texture mapping imaging are equally revealing. The back face of the mechanism carries a detailed text describing the eclipse predictions the device was capable of generating, including glyphs indicating whether a predicted eclipse would be solar or lunar and in some cases what kind of eclipse to expect. The front face carries inscriptions describing the planetary positions the mechanism tracked. A separate section of the text, the so-called Parapegma, coordinated the mechanical calendar with the rising and setting of named stars across the year, a device with deep roots in Greek agricultural and navigational practice. Read together, these inscriptions suggest that the mechanism was not merely a mathematical curiosity but a working instrument intended to be used by someone who needed accurate astronomical information for practical purposes, whether for calendrical coordination, navigation, religious ceremony, or astrological calculation.

Who built it, and where, remains a matter of productive debate. The astronomical knowledge embedded in the device is consistent with the scientific culture of Rhodes in the second and first centuries BCE, and the ancient sources mention Posidonius, a Stoic philosopher and polymath who ran a school on Rhodes around 100 BCE, as someone who built mechanical models of celestial motion. Cicero, writing in the first century BCE, describes seeing a mechanical orrery built by Archimedes, another name that appears in ancient descriptions of mechanical devices of considerable sophistication. Whether the Antikythera Mechanism was built in Rhodes, in Alexandria, in Corinth, or somewhere else, whether it was made by a single craftsman working alone or produced in a workshop tradition with multiple skilled hands, none of these questions has a definitive answer. What can be said is that it was built by someone embedded in a tradition of serious astronomical and mechanical knowledge, and that the level of skill it represents did not appear from nowhere. Somewhere, probably in a workshop with good tools and a strong light, a craftsman sat for months cutting those gear teeth with a precision that challenges modern engineers to explain.

The ship that carried it was likely a Roman vessel sailing from somewhere in the eastern

Mediterranean toward Italy, probably around 60 BCE based on the numismatic and ceramic evidence found alongside it. Whether the mechanism was cargo, personal property of a passenger, a diplomatic gift, or something else is unknown. Whether the instrument was ever used for the purposes it was designed for, or whether it sank on its first voyage, is unknown. The wreck yielded many beautiful objects, but only the mechanism has continued to reshape understanding of the ancient world in the way it has, because it is not merely an object of aesthetic quality but evidence of a whole category of technical capability that was not supposed to have existed.

This is the deeper significance of the Antikythera Mechanism, and it is why the device functions as a useful entry point to the larger argument of this book. It is not simply a remarkable object. It is a corrective. It reveals that the history of technology as it was written through most of the twentieth century contained a significant blind spot, a tendency to assume that because ancient civilizations lacked certain theoretical frameworks or industrial capabilities, they therefore lacked sophistication of all kinds. The mechanism demonstrates that ancient Greek mechanical science was capable of achievements well beyond what surviving texts and other physical evidence had suggested. And it raises an obvious and uncomfortable question that echoes through every chapter of this book: if this one object happened to survive in the Aegean seafloor long enough to be discovered by sponge divers in 1900, what else existed that did not survive?

Serious scholars have proposed that the Antikythera Mechanism was not a unique creation but representative of a broader tradition of mechanical astronomy in the Hellenistic world, a tradition whose other products have simply not survived. The ancient sources contain tantalizing references to mechanisms and devices that sound, in retrospect, like they might belong to this tradition. The gears and the astronomical knowledge they encode were not impossible innovations without precedent; they were built on a foundation of Greek mathematical astronomy that we know existed, using metalworking techniques that were demonstrably available. The anomaly is not that one craftsman was capable of building something like the Antikythera Mechanism. The anomaly is that we have exactly one physical example, where the breadth and depth of the underlying tradition suggests there should have been many more.

What happened to the rest is the story of most ancient technical achievement: the fragility of complex objects, the disruption of the workshops and patronage networks that produced them, the deterioration of bronze and wood and organic materials over millennia, and the failure of the ancient world to preserve, in forms that could survive two thousand years of political upheaval and material decay, the full record of what it had accomplished. The Antikythera Mechanism survived because it went into the sea inside a wooden box, insulated from the processes that destroy bronze on land, and was found before those maritime processes had entirely finished their work. It is a survivor by accident. The rest, whatever it was, is gone.

And yet the mechanism itself endures, in the National Archaeological Museum in Athens, still being studied, still yielding information to new imaging technologies, still capable of producing scholarly surprises. In 2021 a research team at University College London published a new reconstruction of the front plate of the device incorporating the positions of the five planets known to the ancient Greeks, a reconstruction based on newly legible inscriptions that suggested the mechanism's planetary display was more complete than previously understood. The device continues to expand, in the imagination of researchers, into something larger and more elaborate than any single generation of scholars has been able to see in it. After more than a century of study, it remains the most eloquent single object for the argument that the ancient world was doing things we have only begun to understand.

Two

Roman Concrete and the Secret of Eternal Structures

Some of the strongest structures on the planet were built two thousand years ago, are sitting in the sea right now, and are still getting stronger. This is not a poetic exaggeration. It is a statement of materials science, confirmed by laboratory analysis, and it tells us something important about what Roman engineers understood that the modern construction industry spent most of the twentieth century ignoring.

Stand at the entrance to the harbour of Caesarea Maritima, on the coast of what is now Israel, and you are looking at the ruins of one of the most ambitious engineering projects of the ancient world. Herod the Great built this harbour in the late first century BCE, a deep-water port in a place where the natural geography offered nothing of the kind, forcing his engineers to construct artificial breakwaters extending hundreds of metres into the open Mediterranean. They did it using concrete, poured into wooden forms in water up to six fathoms deep, and they used a specific formulation of concrete that has confounded modern engineers and materials scientists for decades: a mixture incorporating volcanic ash from the Italian region of Pozzuoli, roughly two thousand kilometres away, combined with seawater and lime in proportions that ancient Roman builders had worked out through generations of empirical practice. More than two thousand years later, substantial sections of those Caesarea breakwaters are still standing in the sea.

The longevity of Roman concrete is not a myth or an exaggeration. It is a documented materials science phenomenon that has attracted serious academic and engineering attention, particularly since a landmark 2017 study by a research team led by Marie Jackson at the University of California Berkeley, working in collaboration with colleagues from the Lawrence Berkeley National Laboratory and several European institutions. Using X-ray microdiffraction and other analytical techniques unavailable to previous generations of researchers, Jackson's team examined cores of Roman marine concrete taken from ancient harbour structures and compared what they found with the behaviour of modern Portland cement concrete in similar marine environments. The results were striking enough to attract coverage well beyond the usual academic channels.

What the analysis revealed was that Roman marine concrete, rather than degrading when seawater permeated its structure, as modern Portland cement concrete does, was instead undergoing a slow chemical reaction that continuously reinforced and consolidated the material. Seawater penetrating the pores of the concrete was reacting with the aluminous tobermorite crystals forming within the volcanic ash component, producing new mineral growth that filled fractures, bonded aggregate particles more tightly together, and increased the overall density and strength of the material over time. The concrete was, in a measurable sense, self-healing.

Cracks that formed as the structure settled or flexed under wave action were being slowly filled by mineral deposits precipitated from the seawater itself.

Portland cement concrete, by contrast, reacts with seawater in ways that cause the calcium silicate hydrate binding matrix to leach out over time, weakening the material and eventually leading to the spalling, cracking, and structural degradation visible in virtually every modern seawall within a few decades of construction. The comparison is stark: ancient structures improving with seawater exposure, modern ones dissolving under the same conditions. That reversal is the central discovery, and it demands an explanation.

The key ingredient in the Roman formulation was the volcanic ash, which the Romans called pozzolana and which their builders had been using since at least the second century BCE. The best quality came from deposits around the Bay of Naples, particularly from the region around Pozzuoli, the ancient city of Puteoli, one of the most important commercial ports of the Roman world. This ash, produced by the volcanic activity of the Campi Flegrei complex, contains a high proportion of aluminium and silicon in a glassy, reactive form, the legacy of explosive volcanic eruptions that transformed silicate rock into fine particulate material with a large surface area and high chemical reactivity. When mixed with lime and water, this ash underwent a pozzolanic reaction, combining with the calcium hydroxide produced when lime hydrates to form stable calcium silicate and calcium aluminate hydrate compounds. The resulting binder was not merely strong; it was chemically different from any concrete produced without the volcanic component, and its behaviour in contact with seawater was unlike anything else the ancient or medieval world had available.

Roman builders understood this empirically long before any of it was expressible in chemical terms. The first-century BCE architect and engineer Vitruvius, whose ten-volume work De

Architectura remains one of the most important technical documents to survive from the ancient world, described the properties of pozzolana with considerable precision and recommended it specifically for marine and underwater construction. He noted that when volcanic ash from the Bay of Naples region was mixed with lime and rubble and compacted into wooden forms submerged in seawater, the mixture hardened into something as durable as stone. He did not explain why this happened, because he had no chemical theory available to frame the explanation, but he knew that it happened consistently and reliably, and he passed that knowledge on in the form of practical instructions that Roman engineers followed across the empire wherever they could obtain the necessary materials.

The scale on which Roman concrete was used is difficult to grasp without spending time in the presence of the surviving structures. The Pantheon in Rome, completed in its current form under Hadrian around 125 CE, has a dome of unreinforced concrete with a diameter of 43.3 metres that remains the largest unreinforced concrete dome in the world, standing and continuously used as a place of worship for nearly nineteen centuries. The Colosseum was built primarily of concrete and brick, its massive foundations and vaulted corridors poured in a material whose performance under sustained load over two millennia has been studied by structural engineers with something approaching professional envy.

Trajan's Markets, a complex of commercial and administrative spaces sometimes called the world's first shopping mall, used concrete vaulting on a scale that would have been structurally impossible with any other building material available in the ancient world. At Baiae, on the Bay of Naples, the ruins of elaborate bath complexes built directly into and over the sea demonstrate concrete construction in marine environments whose durability even the engineers who built them might not have anticipated. Picture what a

Roman construction site at its height looked like: teams of workers hauling timber formwork into position in the shallows, others mixing volcanic ash with lime by hand in great wooden troughs, the sharp smell of quickslaking lime in the air, foremen moving between gangs calling out adjustments, and the whole enterprise coordinated by the institutional memory of a construction industry that had been refining this method for generations. It was not magic. It was disciplined, experienced, organized labor working with materials they understood deeply.

The Roman concrete tradition was not monolithic. Different applications required different formulations, and Roman builders were pragmatic experimenters who adjusted their mixtures based on what was available locally and what the specific structure demanded. For underwater work, pozzolanic concrete was essential. For vaults and domes, the Romans used increasingly lightweight aggregates as they went higher, incorporating pumice, volcanic tuff, and even empty amphorae in the upper sections of dome structures to reduce weight without compromising the compressive strength of the lower portions. For foundations, they poured massive rubble concrete, using larger aggregate and relying on mass and compressive strength rather than tensile properties. The Pantheon's foundations are estimated to be approximately seven metres deep and seven metres wide, poured in concrete so dense that the structure has shown virtually no differential settlement in two thousand years of occupation.

This is a point that deserves careful handling, because it is easy to misstate. Roman concrete was not universally superior to modern construction material. Portland cement concrete has significant advantages in many applications: it sets quickly, it can be produced with consistent quality at industrial scale from widely available ingredients, it works well in a wide range of non-marine environments, and it can be reinforced with steel to give it tensile strength that unreinforced concrete, Roman or modern, entirely lacks. The Roman building tradition had no equivalent of reinforced concrete, which is why Roman builders relied heavily on arches, vaults, and domes to handle structural loads in ways that kept the material in compression rather than tension.

What Roman marine concrete does better than modern Portland cement in seawater is specifically the long-term durability question in that particular environment, and the reason it does it better is that the pozzolanic chemistry was optimized, through centuries of empirical trial and refinement, for exactly those conditions. That is not a small distinction. Marine construction is among the most demanding structural environments on earth, and the material the Romans developed for it remains, in critical respects, the best ever produced for the purpose.

The question of why the Roman concrete tradition did not survive the fall of the empire is more complex than it might initially appear. It is not simply that the knowledge was kept secret, or that it was lost in a single catastrophic event, or that the formula was buried in a destroyed library. Vitruvius's De Architectura survived the medieval period and was widely read by Renaissance architects; the pozzolanic principle was never entirely forgotten. The problem was more practical and more systemic. High-quality pozzolana came from specific deposits in the Bay of Naples region, and the Roman distribution network that moved volcanic ash from Pozzuoli to construction sites across the Mediterranean depended on an integrated commercial infrastructure: merchant shipping, established trade routes, procurement networks, and the economic conditions of a unified empire that made long-distance transport of heavy bulk materials economically viable.

When the western Roman empire contracted and fragmented in the fifth century CE, that infrastructure collapsed with it. Builders in former Roman territories could still read Vitruvius, but they could not necessarily obtain the volcanic ash he described, and the local alternatives were rarely as reactive and rarely available in the quantities that Roman concrete construction had required. The knowledge survived on the page. The conditions that made the knowledge useful, a functioning state, a reliable labour supply, an organized construction industry with accumulated craft memory, an economic environment in which monumental building was a priority worth enormous investment, all of these vanished together.

The rediscovery of what Roman marine concrete was actually doing chemically is very recent. Jackson's 2017 study and the subsequent work it prompted have inspired serious engineering research into whether pozzolanic concrete formulations might have practical applications in modern marine construction, particularly given the environmental costs of Portland cement production, which requires heating limestone to very high temperatures and produces significant carbon dioxide emissions.

Pozzolanic concrete can be made with lower energy inputs from volcanic materials already processed by geological forces. Whether a modern iteration of Roman-style marine concrete could be produced at industrial scale, and whether its long-term performance would match what the ancient examples demonstrate, remain active questions. The research is ongoing and the engineering interest is genuine.

What the Roman concrete story illustrates, with particular clarity, is the relationship between empirical knowledge and theoretical understanding. The Romans knew what worked. They had tested their formulations over generations of construction practice, observed the long-term behaviour of their materials in real structures under real conditions, and codified what they learned in practical guides transmitted through the construction industry of the empire. What they did not have was any theoretical understanding of why the volcanic ash worked as it did, what chemical reactions were occurring at the molecular scale, or how those reactions related to the durability they were producing.

The theoretical understanding came two thousand years later, when materials scientists with electron microscopes and synchrotron X-ray facilities looked at ancient concrete and figured out the chemistry from the other end. Practical knowledge preceded theoretical knowledge by two millennia. And in the gap between them, the tradition was partially lost, not entirely, but enough that the world spent roughly fifteen centuries building with inferior materials for marine applications because the conditions that had made Roman concrete construction possible as an organized industry no longer existed. That is a genuinely uncomfortable lesson, and it applies far beyond the question of seawalls and volcanic ash.

The harbour at Caesarea Maritima continues to erode, slowly, as the sea and the centuries do their work on what Herod's engineers built two thousand years ago. But it erodes far more slowly than the modern concrete structures built nearby in the twentieth century, which are already showing the typical signs of Portland cement degradation in a marine environment. Visitors who know what they are looking at find this gap deeply thought-provoking: the ancient structure outlasting the modern one not because of any mysterious property, but because the people who built it had spent generations learning, through careful practical observation and accumulated trial and error, exactly what their particular environment required and how to give it that.

Three

Damascus Steel and the Lost Art of the Perfect Blade

In the markets of the Levant, a European crusader could buy a sword that his own smiths could not make. He could hold the blade up to the light and watch the surface ripple with a pattern like water moving over sand, flowing bands of crystalline structure visible beneath the polished steel, shifting as the angle changed. He could test the edge on a hair balanced across the blade and watch it part without resistance. He could flex the steel and feel a springiness that European ironwork of the period, with its more brittle high-carbon edges and softer iron cores, simply did not possess. He could carry this blade into battle and find that it held its sharpness through combat that would have dulled a sword from Toledo or Cologne. He could not explain any of this. Neither, for several more centuries, could anyone else.

The material that produced this combination of properties had a history stretching back considerably further than the crusading era, and considerably further east than Damascus. Its origins lay in southern India, primarily in the Tamil-speaking regions around the city that the ancient world knew as Kodumanal, and in the workshops of the Deccan plateau, where smiths had developed by at least the early centuries of the common era a method of producing high-carbon steel of exceptional purity and consistency using a small enclosed crucible fired at high temperatures. The steel they produced in these crucibles was called wootz, a name that appears in various English transliterations of the Tamil and Telugu words for the material, and it was traded in the form of small cakes or ingots, typically weighing around a kilogram, that merchants carried along the overland and maritime trade routes connecting India to the Persian world, the Arabian peninsula, Central Asia, and eventually the Mediterranean.

The smiths of the Islamic world, particularly those working in Persia, Syria, and along the trade routes of Central Asia from roughly the ninth century CE onward, learned to work wootz steel in ways that produced the distinctive surface pattern associated with the finest Damascus blades. The name Damascus steel applies both to the finished blades made from wootz in various centres of Islamic metalworking and to the characteristic pattern that appears on their surface when the material is correctly forged, manipulated, and etched. That pattern, technically described as a Widmanstatten structure or a form of carbide banding depending on which aspect of the microstructure one is examining, was the visible result of a complex interplay between the carbon chemistry of wootz steel and the specific techniques of heating, forging, and cooling that the best smiths had mastered through generations of practice.

Understanding what was happening metallurgically in the finest Damascus blades requires a brief engagement with the chemistry of iron and carbon, because the properties of steel are almost entirely determined by the ways in which carbon atoms are distributed within the iron crystal lattice. Pure iron is soft and relatively ductile, easily worked but unable to hold a sharp edge and easily bent under load. Adding carbon to iron changes everything: the carbon atoms disrupt the iron crystal lattice in ways that inhibit the movement of dislocations, the microscopic defects in crystal structure through which metals deform, and this disruption is what gives steel its hardness and strength.

But carbon content is a double-edged property. Too little, and the steel is too soft. Too much, and it becomes brittle, hard enough to hold an extraordinary edge but liable to fracture catastrophically under impact. The range within which steel is simultaneously hard, strong, and tough enough not to shatter is relatively narrow, and the art of the steelmaker lies in achieving the right carbon content and, critically, in controlling how the carbon is distributed through the material. This is the problem that wootz, and the techniques developed to work it, solved in ways that Western metallurgy would not approach for centuries.

Wootz steel was produced with a carbon content typically in the range of 1.0 to 1.8 percent, which is technically in the category of high-carbon or ultra-high-carbon steel, well above the carbon content of most Western medieval steels. At these carbon levels, the steel contains, in addition to the carbon dissolved within the iron crystal lattice, a separate population of iron carbide particles, a compound called cementite, distributed through the material. In ordinary high-carbon steel, produced by rapid heating, forging, and cooling, these carbide particles form a relatively disordered, fine-grained structure. In wootz steel, produced by the slow cooling of a molten charge within a sealed crucible and then carefully worked by a smith who understood the material's thermal behaviour, the carbide particles organized themselves into bands and networks aligned with the forging direction, creating the characteristic layered microstructure that became visible, when the surface of the finished blade was etched with mild acid, as the flowing patterns that gave Damascus blades their unmistakable appearance.

The visual pattern was not merely decorative. The banded carbide structure had mechanical consequences: the alternating regions of higher and lower carbide concentration gave the blade a subtle composite quality, with the carbide-rich bands contributing hardness and the carbide-lean bands contributing toughness, creating a material that performed better across the full range of sword use than a homogeneous steel of the same average carbon content could have done. Medieval Islamic metallurgical writers were aware of this connection and described the best blades specifically in terms of the quality and regularity of their surface patterns, understanding at a practical level, if not in the chemical vocabulary available to modern materials scientists, that the pattern and the performance were related.

The finest blades were not simply cast from wootz and etched. They were the product of extensive forging work that developed the carbide banding in controlled ways, work requiring intimate knowledge of the temperature range within which the material behaved correctly and the rate at which it needed to be cooled between heats. A master smith at his forge knew the correct temperature not from any measuring instrument but from the colour of the glowing steel itself: the dull red of iron that is too cool to work without cracking, the bright cherry of a temperature at which the metal begins to move freely under the hammer, the pale straw yellow near the upper limit where the carbide network, built up through careful repeated heating, begins to dissolve back into the matrix and the hours of work are undone.

He knew how many hammer blows he could deliver at a given heat before the metal dropped below the working range, how the sound of the steel on the anvil changed as it cooled, and what the faint sheen on the surface in certain lights told him about the state of the carbide structure beneath. None of this was written down in any form that could convey it to someone without years of experience at the forge. It was knowledge that lived in the body, accumulated through thousands of hours of practice, and it was exactly the kind of knowledge that vanishes when the workshops that carry it are disrupted.

The question of when and why the production of high-quality Damascus steel declined has been debated by historians of technology since the subject attracted serious scholarly attention in the nineteenth and twentieth centuries. The most commonly cited date for the effective end of the tradition is somewhere in the early to mid-eighteenth century, though the decline was gradual rather than sudden and is difficult to pin to a single cause. The supply of Indian wootz steel ingots, which was the essential raw material for all genuine Damascus blade production, became increasingly disrupted from the late seventeenth century onward as the Mughal empire fragmented and the commercial networks that had moved wootz along overland routes from the Indian interior to the workshops of Persia and Syria became less reliable and eventually ceased to function at the scale the trade had previously sustained.

There is also a materials science dimension to the supply problem that only became clear through modern analytical work. The specific microstructure of the finest Damascus blades appears to depend not only on the carbon content of the wootz but on the presence of trace quantities of specific impurities: vanadium, chromium, manganese, molybdenum, and other elements present in tiny amounts in certain Indian ore sources but not in others. These trace elements influenced the formation of the carbide network during solidification and forging, nudging the chemistry in directions that produced the finest patterns and the best cutting properties. Smiths working with wootz from ore sources that happened to contain the right trace elements could achieve results that smiths working with superficially similar wootz from different sources could not replicate, even using identical techniques.

The geographical specificity of the best material was a structural vulnerability in the tradition: when those particular ore sources became inaccessible or were simply exhausted, the quality declined in ways that no amount of skill could compensate for. The best Damascus blades were not just the product of superior knowledge; they were the product of a specific alignment between a geological accident of ore chemistry, a long-distance trade network, and a craft tradition capable of exploiting both. Remove any element and the whole combination fails. This is the kind of fragility that historical accident can expose without warning.

The modern effort to understand and replicate Damascus steel has been one of the sustained intellectual exercises in the history of materials science, drawing in researchers from metallurgy, archaeology, physics, and chemistry over the better part of two centuries. The Sheffield cutler and metallurgist John Stodart, working with the chemist Michael Faraday in the early nineteenth century, made some of the first serious attempts to analyze and reproduce the material, with limited success. Subsequent researchers across Europe and America produced numerous partial reconstructions and competing theories. The problem was that the characteristic surface pattern could be reproduced by various techniques, including the folded and welded construction of Japanese sword blades, which produces a somewhat similar visual effect through entirely different metallurgical means, and distinguishing genuine reconstructions from superficially similar imitations required levels of analytical precision that earlier researchers did not always have available.

The most consequential modern discovery came in 2006, when the German researchers Peter Paufler, Peter Berger, and their colleagues published a paper in the journal Nature reporting that analysis of a genuine historical Damascus blade, using high-resolution transmission electron microscopy, had revealed the presence of carbon nanotubes within the cementite network of the steel. Carbon nanotubes are cylinders of carbon atoms arranged in a hexagonal lattice and rolled into a tubular shape, a structure first described and characterized by modern nanotechnology researchers in the early 1990s and widely considered one of the most exciting discoveries in materials science for their extraordinary mechanical and electronic properties. Finding them in an eighteenth-century sword blade was not simply surprising; it was a demonstration that natural thermal processing under certain specific conditions could produce nanotube structures spontaneously, without any of the controlled deposition techniques that modern nanotechnology uses to create them deliberately.

Whether the nanotubes contributed to the cutting properties of Damascus blades, or whether they were merely an incidental microstructural feature produced alongside the carbide network that actually mattered for performance, remains a subject of ongoing research and debate. What the finding established was that the microstructure of the finest Damascus blades contained features that were not identified or understood until two decades into the twenty-first century and that no smith working in the seventeenth or eighteenth century could possibly have known were there. The craftsmen were producing nanotube-reinforced steel without knowing it, guided entirely by accumulated empirical knowledge of how the material behaved under specific conditions of heat, impact, and cooling. The theoretical explanation came three hundred years after the craft tradition that produced it had collapsed.

Today, researchers in several countries continue working on the problem of reproducing Damascus steel at the level of the best historical examples. Some groups have achieved materials with similar visual patterns and broadly comparable mechanical properties. Others have focused on the specific chemical conditions required for nanotube formation and the role of trace elements from historical ore sources. The consensus, such as it is, is that a reasonable approximation of traditional Damascus steel can now be produced by researchers who understand the chemistry, but that the finest historical examples remain difficult or impossible to replicate exactly, in part because the specific ore sources are gone, in part because the full operational knowledge of the best smiths was never captured in any text, and in part because the relationship between trace element composition, thermal processing, and final microstructure is complex enough that even modern analytical chemistry has not yet reduced it to a simple reproducible procedure.

The story of Damascus steel is, in miniature, a story about the relationships between materials, knowledge, trade, and time. A specific combination of geological accident, long-distance commerce, and generational craft knowledge produced blades that metallurgists with electron microscopes and modern theoretical tools are still working to fully understand. When any one element of that combination was disrupted, the whole thing unravelled. The blades that survive in museum collections are evidence not only of what the smiths who made them could do, but of how specific, how contingent, and how irreplaceable the combination of conditions that made them possible actually was. In that sense, every surviving Damascus blade is not just a weapon or an object of beauty. It is a record of something the world learned once, practiced for centuries, and then lost.

Four

Greek Fire and the Weapon That Burned on Water

The sailors who first saw it coming described something that seemed to belong to a different order of warfare entirely. A Byzantine naval vessel would draw close, and from somewhere along its bow a jet of fire would erupt, not an arc of burning debris thrown by a catapult, not a torch or a burning projectile of the kind that siege engineers had been producing for centuries, but a sustained stream of liquid flame that traveled across the water and continued to burn after it landed. It burned on the surface of the sea. It could not be extinguished with water; the attempts to smother it with water may have made it worse. It clung to wood, to rigging, to cloth, to human skin, and it kept burning as men leapt into the sea trying to escape it. The noise was described as a continuous thunder. The light was described as blinding. Arab sailors who encountered it in the late seventh century wrote about the experience with a terror that reads, even across more than a thousand years, as entirely sincere.

Greek Fire, as it came to be known to the Western world, though the Byzantines themselves used various terms including liquid fire, sea fire, and Roman fire, was a military technology developed by the Byzantine Empire in the latter half of the seventh century CE, at a moment when the empire's survival depended on controlling the sea lanes around Constantinople. The Arab fleets that besieged the Byzantine capital in 674 CE and again in 717 CE were among the largest naval forces assembled in the ancient or early medieval world, and the Byzantines, despite their engineering sophistication and their disciplined military tradition, were significantly outnumbered. What they had that the Arab fleets did not was Greek Fire, and it proved decisive. Both sieges were broken. The Arab fleets, in both cases, were destroyed or driven back with casualties that the sources describe in terms suggesting near-total annihilation of the attacking force. The weapon had, in its first major deployments, demonstrated a battlefield effectiveness that would make it the most feared naval weapon in the medieval Mediterranean for the better part of five centuries.

The historical attribution of the weapon's invention is almost universally given to a man named Kallinikos, a Syrian architect or engineer who is said to have fled to Constantinople from the Arab-controlled city of Heliopolis, in what is now Lebanon, sometime around 668 CE. The Byzantine chroniclers credit Kallinikos with bringing a new incendiary formula to the imperial court, and the timing is consistent with the first documented use of the weapon in the naval battles of the 670s. Whether Kallinikos genuinely invented the formulation from scratch, or whether he refined and weaponized an existing incendiary tradition, or whether the Byzantine sources are attributing to a single individual what was in fact a more collective development, is impossible to say with certainty from the surviving record. What is clear is that the deployment of the weapon under the early

Macedonian emperors was deliberate, systematic, and coordinated with purpose-built delivery technology.

The delivery system is one of the aspects of Greek Fire about which the historical sources are most informative, and it was clearly as important as the chemical formulation itself. Byzantine warships equipped with the weapon carried bronze or iron tubes, called siphons, mounted at the bow or sometimes at the sides of the vessel, through which the liquid could be projected under pressure. The pressurization mechanism is not described in surviving sources with enough precision to reconstruct it definitively, but the most plausible interpretations suggest some form of bellows or pump system, possibly incorporating heated air or steam pressure to force the liquid through the siphon with sufficient velocity to project it as a jet rather than a dripping stream. Byzantine manuscript illustrations, several of which survive from the tenth and eleventh centuries, show the siphon tubes mounted on warship prows, with what appear to be pump mechanisms visible in the hull, and they show the fire projected in a sustained beam rather than in a lobbed arc.

There were also hand-thrown versions of the weapon, smaller containers of the incendiary material that could be hurled by soldiers or dropped from walls onto besieging forces below, and catapult-launched versions using clay pots or similar vessels as delivery containers. The naval siphon, however, was the form that defined the weapon's military significance. It turned a Byzantine warship into something that had no precedent in ancient naval warfare: a vessel capable of projecting sustained, directed flame at an enemy ship while maintaining the speed and maneuverability to close with that ship and retreat before a counterattack could be organized. The combination of the weapon and the fast dromons, the light Byzantine warships purpose-built for speed, created a naval capability that the Arab fleets, for all their numerical superiority, could not match.

Picture the crew responsible for the siphon in the moments before an engagement. They worked in the narrow bow of a fast dromon, crouched around a bronze tube connected by copper piping to a sealed reservoir below decks. The heat from the pressurizing mechanism would have been considerable, the smell of the naphtha sharp and persistent even before ignition, and the noise of the pump competing with the sound of oars and the shouts of the captain calling the closing distance. When the order came, one man worked the pump, another managed the direction of the tube, and the officer watching the gap to the enemy vessel understood that the range had to be exact before the fire was released. Too far and the jet fell short or dispersed into spray. Too close and the dromon itself risked the reflected heat. The crews who operated the siphon trained for this role specifically, and the weapon's battlefield effectiveness depended on their discipline as much as on the chemistry of the mixture they were projecting.

The question that has fascinated historians, chemists, and military historians for centuries is the formulation itself: what was Greek Fire actually made of? The Byzantines guarded the secret with extraordinary care, and the secrecy was not merely institutional habit. The emperors of Constantinople regarded the weapon's formula as a state secret of the highest order, restricting knowledge of its preparation to a small number of trusted specialists and threatening severe punishment for any disclosure. Constantine VII Porphyrogennetos, writing in the tenth century in a handbook of imperial governance intended for his son, described Greek Fire as one of three things that must never be revealed to foreigners under any circumstances, alongside the method of manufacturing the imperial purple dye and certain ceremonial protocols of the court. He explicitly attributed the formula to divine revelation, writing that an angel had revealed it to Constantine the

Great, a claim that was almost certainly political mythology rather than theological assertion, but that indicates how jealously the Byzantines protected the knowledge and how central it had become to their imperial identity.

Despite the secrecy, various enemies of the empire attempted to steal, reverse-engineer, or independently develop the weapon. Some Arab sources from the eighth and ninth centuries describe incendiary mixtures that may reflect partial knowledge of the Byzantine formulation, or may reflect independent development of related but distinct incendiary technologies. There are references in later medieval sources to attempts by Crusaders, Russians, and various other parties to acquire the formula, and a handful of accounts claim that the secret was partially transmitted at various points. None of these accounts is reliable enough to reconstruct the actual formulation, and the modern debate about what Greek Fire contained has proceeded largely on the basis of physical and chemical reasoning from the weapon's described properties.

The Byzantine chronicler Theophanes, writing in the early ninth century, provides one of the most detailed surviving accounts of the weapon's early deployment, describing the Byzantine ships advancing with fire-throwing tubes at their prows and the flames burning on the water in ways no enemy action could suppress. He was drawing on sources close to the events themselves, and his account has the specificity of a writer who understood that what he was describing was militarily decisive rather than merely dramatic. Later, Anna Komnene, writing her history of her father Alexios I in the twelfth century, referred to the weapon with enough technical confidence to make clear that it remained in active Byzantine service and that its basic operating principle, if not its precise formula, was understood by educated observers of her era. These chroniclers do not supply a recipe. What they supply is confirmation, from within the Byzantine literary tradition itself, that the weapon was real, consistent, strategically significant, and deliberately protected from the knowledge of enemies across centuries of active use.

Those described properties, compiled from the various Byzantine and Arab sources that mention the weapon's behavior, form a reasonably consistent picture. The substance was liquid at room temperature or moderately warm temperatures and could be stored in sealed containers for transport and use at sea. It ignited on contact with water, or at minimum burned intensely in contact with water rather than being extinguished by it. It adhered strongly to surfaces including wet wood and human skin. It could be projected under pressure through a tube as a sustained jet. It burned with great intensity and produced thick, dark smoke. Some sources describe it as having a foul or sulfurous smell. And it was produced in a form stable enough to be stored in quantity aboard naval vessels without spontaneous ignition, which imposes significant constraints on any proposed formulation involving highly reactive ingredients.

The modern candidates for the active components of Greek Fire fall into several broad categories, and most serious analyses of the problem conclude that the weapon was likely a complex mixture rather than a single substance. Crude petroleum or naphtha, available in the region around the Black Sea and in parts of the Middle East, is almost universally included in proposed formulations, and is probably the most important single ingredient. Naphtha burns intensely, is liquid at room temperature, can be pressurized and projected through a tube, and floats on water rather than mixing with it, which would explain the weapon's ability to burn on the sea surface. Some versions of the proposed formulation add quicklime, calcium oxide, which reacts vigorously and exothermically with water, potentially explaining the accounts of the weapon igniting when water was applied to it. Rosin, pine tar, and various other organic compounds that are highly flammable and adhesive have also been proposed as thickening agents that would give the mixture the clinging quality described by the sources.

Some researchers have proposed that the key innovation of Kallinikos may have been the addition of a specific reactive component that caused the mixture to auto-ignite when heated or pressurized through the siphon delivery system, which would explain why the weapon appeared to ignite spontaneously at the nozzle without the need for an open flame at the point of projection. Phosphorus compounds, which auto-ignite on contact with air, have been proposed, though the availability of purified phosphorus in the seventh-century Mediterranean is questionable. Heated naphtha itself will auto-ignite if brought to a sufficiently high temperature before projection, and the pump mechanism of the siphon may have incorporated a heating element that served this function. The truth is that no single proposed formulation fully accounts for all the described properties, and the honest position is that we do not know exactly what Greek Fire contained.

What we do know is that the weapon worked, spectacularly and repeatedly, for a period of several centuries, and that it was decisive in at least two engagements that shaped the course of Mediterranean history. The defeat of the Arab siege fleets in the 670s and 717 to 718 CE preserved Byzantine control of Constantinople at a moment when the Arab expansion was moving with extraordinary speed across the Mediterranean world. Had Constantinople fallen in either of those sieges, the subsequent history of southeastern Europe and the Aegean would have been fundamentally different. Greek Fire did not win those battles alone, but the historical consensus is that it provided the margin without which the Byzantine fleet could not have held the sea against numerically superior forces. It is one of relatively few cases in military history where a single technology can be identified with reasonable confidence as having altered the strategic balance at a genuinely pivotal moment.

The weapon remained in Byzantine service through the tenth and eleventh centuries, and there are accounts of its use in engagements against the Rus, against Bulgarian naval forces, and in various other conflicts of the period. Gradually, however, references to it in the sources become less frequent, and by the time of the Crusades the Byzantine fleet no longer appears to have deployed Greek Fire with the devastating effect it had produced in earlier centuries. Whether the formula had been lost, whether the production infrastructure had been disrupted, whether the weapon had been superseded by other tactical developments, or whether the relevant source material simply reflects a gap in documentation rather than a genuine disappearance of capability, scholars debate without resolution. The fall of Constantinople to the Fourth Crusade in 1204 CE, the dismemberment of the Byzantine state, and the subsequent political fragmentation of what had been the empire's administrative and military apparatus provide plausible historical conditions for the loss of a technology whose production required specialist knowledge, controlled ingredients, and an organized military-industrial support system.

By the time Byzantine-style naval power had ceased to exist as a meaningful strategic entity, the specific technical knowledge that had produced Greek Fire appears to have gone with it. Later medieval and early modern incendiary weapons used by various European, Arab, and Ottoman forces show no evidence of the specific combination of properties that made Byzantine Greek Fire distinctive. The term itself survived in various forms, appearing in medieval military manuals and descriptions of incendiary technology, but the substances described under that name in later sources bear little resemblance to the weapon deployed at Constantinople and seem to reflect the ordinary incendiary tradition of burning oils and resins rather than anything more sophisticated. The secret that Constantine VII Porphyrogennetos had described as divinely given and permanently protected had, in the end, been lost along with the state whose identity it had helped define.

What remains instructive about Greek Fire, beyond the military and chemical history, is what it reveals about the relationship between technology and the political structures that sustain it. The Byzantines did not simply invent an incendiary mixture; they built around it an entire system of specialized production, controlled distribution, purpose-built delivery mechanisms, trained operators, tactical doctrine, and institutional secrecy that together constituted the weapon as a military capability. Any one element of that system in isolation would have been insufficient. The formula without the siphons, the siphons without the trained crews, the trained crews without the tactical doctrine, the tactical doctrine without the institutional secrecy that prevented the formula from spreading to enemies, none of these alone constituted Greek Fire. The weapon was the system, and when the empire that maintained the system collapsed, the weapon died with it, exactly as one would expect of a technology whose power depended not on a written recipe but on the integrated functioning of a complex military-industrial organization.

Five

The Baghdad Battery and the Mystery of Ancient Electricity

In the spring of 1938, a German archaeologist named Wilhelm Konig was working through the storage collections of the National Museum of Iraq in Baghdad when he came across a group of clay vessels that had been excavated some years earlier from a site at Khujut Rabu'a, near the village of Khuyut Rabbou'a southeast of the city. The objects were modest in appearance: terracotta jars roughly fourteen centimetres tall, each sealed at the top with a bitumen stopper through which a copper cylinder had been inserted, and within the copper cylinder an iron rod. The jars dated to the Parthian period, roughly between 250 BCE and 224 CE, making them somewhere between seventeen hundred and twenty-two hundred years old at the time of Konig's examination. He published a short paper in 1940 proposing that the objects were galvanic cells, batteries in the modern sense, capable of generating a small electrical current if the copper and iron components were connected and the vessel filled with an acidic liquid such as vinegar or grape juice.

Konig's interpretation was immediately striking, and it has never entirely gone away. The idea that ancient peoples in the Parthian world were generating electrical current more than a millennium before Volta's experiments in the 1790s, which produced the first modern battery, has an obvious and powerful appeal. It has been repeated in countless popular science books, television documentaries, and general histories of technology, often without the qualifications that the serious scholarly debate about these objects demands. The Baghdad Battery, as the objects are almost universally known in popular discussion despite the fact that there are several of them and they were found near rather than in Baghdad, has become one of the most cited examples in the genre of writing that presents the ancient world as having possessed electrical or technological knowledge that modern historians have ignored or suppressed. That popular narrative has made the objects harder to evaluate dispassionately, because the enthusiasm of the proponents and the irritation of the critics have between them generated rather more heat, in both the literal and figurative senses, than the modest clay jars themselves strictly warrant.

The starting point for any honest assessment is the physics, which is entirely straightforward. If you take a copper cylinder, insert an iron rod, place both inside a terracotta vessel, fill the vessel with a mild acid, and connect the copper and iron externally with a conductor, you will generate a small electrical current. This is basic galvanic chemistry: two dissimilar metals in an acidic electrolyte produce a potential difference, and if the circuit is closed, current flows. The experimental record on this point is not in dispute. Willard Gray of the General Electric High Voltage Laboratory built a working replica using grape juice as the electrolyte in 1940, the same year Konig published his interpretation, and measured a potential of approximately 0.5 volts. Subsequent experimenters using various electrolytes have produced between 0.5 and 2 volts from replicas of similar construction. The Khujut Rabu'a objects, if filled with an acidic liquid, would function as batteries. That much is certain.

Willard Gray's 1940 reconstruction is worth pausing on for a moment. He took a replica jar, inserted the copper and iron components, filled it with grape juice, connected the two metals with a wire, and held a galvanometer to the circuit. The needle moved. It was a modest half-volt of direct current, far too weak to power a light or drive a motor, but enough to plate a thin film of metal onto a prepared surface given sufficient time and the right electrolyte solution. Gray sat in his laboratory in 1940 measuring the output of a two-thousand-year-old clay jar and confirmed, quietly and without fanfare, that the object worked exactly as Konig had proposed. What his experiment could not answer, and what no experiment since has been able to answer, was whether anyone in the Parthian world had ever done the same thing deliberately, understood what they were producing, or found a use for it.

The critical question, and the one on which everything else depends, is whether the Parthian craftsmen who made these objects intended them to function as batteries, understood that they produced electrical current, and used them for any purpose that exploited that current. The physical capability to generate electricity does not, by itself, establish intentionality or use. A clay pot filled with acidic liquid and containing iron and copper components could have been made for purposes entirely unrelated to electrochemistry, and the fact that it happens to function as a galvanic cell does not mean its makers knew or cared about that property. The question is whether there is evidence, either in the objects themselves, in the archaeological context in which they were found, or in the written record of the Parthian world, that anyone in this time and place was making, using, or thinking about anything that required an electrical power source.

The archaeological context of the Khujut Rabu'a finds is unfortunately poorly documented by modern standards, because the excavations that produced them were conducted before the systematic recording methods that characterize professional archaeology today had become universal practice. The objects were found in a context that also included scrolls, suggesting a possible connection with some kind of administrative or craft site, but the details of exactly where in the site each object was found, what other objects were associated with them, and what the stratigraphic context indicated about their use or storage are not available with the precision that would allow confident conclusions. This is a genuinely limiting problem, because archaeological context is often the most important evidence available for interpreting ambiguous objects, and in this case that evidence is largely inaccessible.

The physical examination of the surviving objects provides some useful information and some further complications. The bitumen seals that closed the tops of the jars appear to have been functional closures rather than decorative elements, which is consistent with either a battery interpretation or with various storage vessel interpretations. The copper cylinders show some corrosion consistent with acidic exposure, though copper corrodes in many conditions and the presence of corrosion does not establish that an acid electrolyte was specifically used. The iron rods, where they survive, are positioned centrally within the copper cylinders in a way that would maximize the surface area interaction between the two metals, which some commentators have noted is consistent with an intentional electrochemical design, though whether this reflects deliberate optimization or simply the most practical way to insert a rod into a cylinder is impossible to determine. No traces of any electrolyte have been identified in the surviving objects, which may simply reflect the volatility of organic acids over two thousand years rather than indicating that no electrolyte was ever present.

The most commonly proposed use for an ancient battery, if the objects were indeed used as galvanic cells, is electroplating: the deposition of a thin layer of one metal onto another using electrical current. Electroplating requires a power source, an electrolyte solution containing dissolved metal ions, and appropriate electrode arrangements, none of which are technically beyond the capabilities of ancient craftsmen if they understood the relevant processes. There are ancient objects from the Mesopotamian and Egyptian world that have thin metallic coatings whose application is not fully explained by the gilding and plating techniques that are otherwise well documented. A 2003 study by the Italian researcher Ermanno Mariotti and colleagues reported finding ancient artifacts with gold coatings thinner than one micrometre, a thickness that is difficult to achieve by standard ancient fire-gilding methods and that could in principle be consistent with electrodeposition. This is the strongest line of circumstantial evidence for the practical use of ancient galvanic cells, but it is circumstantial rather than direct, and alternative explanations for thin metallic coatings, including various mercury amalgam gilding processes, have not been definitively ruled out.

The skeptical position, held by the majority of academic archaeologists and historians of technology who have engaged with these objects seriously, does not deny the electrochemical capability of the Khujut Rabu'a vessels. It denies the claim that this capability was intentional or that it was ever exploited. The most common alternative interpretations of the objects' original purpose treat them as storage vessels for papyrus scrolls, which fits the archaeological context at the site and the use of bitumen sealing, or as containers for some kind of sacred or ritual substance. The presence of copper and iron in a sealed jar is not, from this perspective, evidence of galvanic design; it is evidence of whatever the jar was actually being used to store, which may have happened to produce galvanic chemistry as an unintentional byproduct or may simply reflect the use of available materials for structural purposes without any electrochemical significance.

Paul Craddock, a metallurgist at the British Museum whose research into ancient metalworking technology is among the most thorough in the field, has examined the question carefully and concluded that while the objects could function as batteries, there is no compelling evidence that they were designed or used as such, and that the absence of any electroplated objects in the Parthian archaeological record attributable to electrodeposition, combined with the absence of any textual reference to electrical phenomena in ancient Mesopotamian, Persian, or Parthian sources, makes the battery interpretation a hypothesis without positive supporting evidence rather than a documented historical fact. This is a reasonable summary of the state of the scholarly debate, and it represents neither dogmatic dismissal of an interesting possibility nor credulous acceptance of a claim that exceeds the available evidence.

What the Baghdad Battery story illustrates more clearly than almost any other example in this book is the epistemological challenge at the center of writing about ancient technology: the difference between what an object could have done and what it was made to do. The ancient world was full of objects whose chemical, physical, or mechanical properties were exploited without any theoretical understanding of why those properties existed, and it is entirely plausible, on general grounds, that ancient craftsmen made discoveries about the behavior of metals in acidic solutions through the ordinary processes of empirical observation. It is also entirely possible that they stumbled upon galvanic effects without recognizing them as such and without developing any systematic use of them. The gap between accidental observation and deliberate technological application is not automatically bridged by physical capability.

The genuinely important question the Baghdad Battery poses is not whether the Parthians had batteries. It is what standard of evidence we require before accepting that an ancient civilization possessed a technical capability that the historical record does not explicitly document. If the answer is that physical possibility is sufficient, then ancient technology becomes an unconstrained field in which almost any capability can be attributed to almost any civilization without evidence. If the answer is that only fully documented, contextually clear, textually supported evidence counts, then some genuine ancient achievements, transmitted through practice rather than text, may be systematically excluded. The honest position is somewhere between these extremes, and navigating it well requires exactly the kind of careful, evidence-graded reasoning that popular writing on ancient mysteries consistently fails to apply.

The Khujut Rabu'a vessels remain in the collections of the Iraq Museum in Baghdad, where they have survived, not without difficulty, the turbulent history of that institution through the late twentieth and early twenty-first centuries. They continue to attract attention from researchers, popular writers, and filmmakers drawn to the idea of ancient electricity, and they continue to be cited in books and documentaries as definitive evidence of something the ancient world was doing with electrochemical technology. The serious scholarly literature is more cautious. What the objects are, beyond terracotta jars with copper and iron components of uncertain purpose, what they were used for, and whether their makers had any understanding of the electrical properties they possessed, remain genuinely open questions. The honest answer is that we do not know. That answer is less satisfying than the one offered by the popular literature, but it is the one that the evidence supports, and in a subject as important and as frequently distorted as this one, intellectual honesty is more valuable than a confident story.

Six

Egyptian Precision and the Problem of Stone

Stand in the King's Chamber of the Great Pyramid at Giza and look at the walls. The granite blocks that line that room, each one weighing between twenty-five and seventy tonnes, fit together at their joints with a precision that a trained stonemason, running a finger along the seam, will find deeply unsettling. The gap at the contact surfaces, in the finest examples, is less than half a millimetre across an expanse of several metres. The surfaces themselves are flat to a degree that, when measured with modern instruments, approaches the tolerances of machine-ground optical components. Those surfaces were not produced by machine. They were produced by human hands, using copper tools, wooden mallets, abrasive sand, and instruments no more technologically sophisticated than a straight edge and a set square, in a culture that had no steel, no electricity, and no theoretical science in the modern sense of the term. What they had was time, organizational capacity, accumulated craft knowledge, and a standard of quality in monumental stone work that the Western world would not surpass and arguably has not yet equalled.

The Great Pyramid is the most famous product of the Egyptian stone-cutting tradition, but it is far from the only one that raises serious questions about how its results were achieved. The statue of Ramesses II at Luxor, the obelisks at Karnak, the casing stones of the pyramid of Khafre still visible at its apex, the granite sarcophagi in the Valley of the Kings, the rock-cut temples of Abu Simbel: across more than three thousand years of monumental construction, Egyptian craftsmen worked stone at a scale and to a standard of precision that continues to attract the attention of engineers and materials scientists who study it with something between admiration and perplexity. Not because it is inexplicable, but because the explanation, when worked out in full, reveals a level of organizational and technical achievement that most people, conditioned by popular assumptions about ancient simplicity, have not adequately accounted for.

The starting point is the material itself. The Egyptians worked several types of stone, each with different properties and different challenges. Limestone, which is relatively soft and easily shaped with copper tools, was used for the bulk of pyramid construction and for the lower courses of many temples. Sandstone, quarried from the cliffs of Upper Egypt and used extensively from the New Kingdom period onward, was somewhat harder but still workable with the tools available.

Granite, quarried primarily from Aswan in the far south, is a different proposition entirely. It is an extremely hard, coarse-grained igneous rock composed largely of quartz, feldspar, and mica, with a hardness that makes it highly resistant to copper tools and that even modern steel struggles to cut efficiently without significant tool wear. The finest internal chambers of the Great Pyramid, including the King's Chamber and the Grand Gallery, are lined with Aswan granite. The sarcophagi, the corner pieces, the plugging blocks that sealed the ascending passages were all granite, all cut and finished to extraordinary tolerances, all moved from a quarry approximately eight hundred kilometres south of Giza and placed with a precision that suggests the quarrying, transport, and fitting operations were coordinated within a single, highly organized system.

The quarrying methods for granite have been studied extensively at Aswan, where the ancient quarry landscape is extraordinarily well preserved, including an unfinished obelisk that was abandoned in place when a crack developed during extraction, giving modern researchers an exceptionally clear view of the techniques in use. The primary quarrying method involved creating a series of holes along the intended line of separation, inserting dry wooden wedges, and then wetting them, causing the wood to expand and drive the stone apart along the prepared line. This technique, simple in principle, required considerable skill in practice: the holes had to be placed at precisely the right spacing and depth, the line had to be prepared in a way that accounted for the natural grain and fracture patterns of the stone, and the sequence of wetting and monitoring had to be managed to ensure that the stone separated along the intended plane rather than cracking unpredictably. The finished surfaces left by this process were rough, and the precision work came after separation.

The tools used for finishing granite and the methods by which the extraordinary surface quality of the finest Egyptian stonework was achieved have been the subject of sustained debate and experimental investigation. The Egyptologist and experimental archaeologist Denys Stocks spent many years working with replica ancient Egyptian tools, cutting granite, limestone, and basalt using copper tools, tubular copper drills, and stone pounders of the type found at ancient quarry sites, and his research provides the most detailed practical account available of what ancient Egyptian stone workers could accomplish with their actual toolkit. His findings are important precisely because they correct two opposing errors that frequently appear in discussion of Egyptian stonework: the error of those who dismiss the achievement as unremarkable because it was done with primitive tools, and the error of those who argue that the precision achieved was impossible without modern or supernatural assistance.

What Stocks demonstrated, through painstaking experimental work, is that copper tubular drills, when used with an abrasive slurry of quartz sand, are effective cutting tools for granite, producing cylindrical cores and flat-bottomed holes of the kind found in ancient Egyptian artifacts. The process is slow by modern standards; cutting a centimetre into granite with a copper drill and sand might take an hour or more of continuous work, and the copper wears rapidly, requiring frequent replacement of the cutting edge. But it works, and it works to a level of precision that is determined not by the tool itself but by the skill and attention of the person operating it. A careful operator, taking time, using good abrasive, and monitoring the drill's tracking constantly, can produce a hole in granite that is straight, smooth, and accurately dimensioned. The speed is not important when the workforce is large and the time available is measured in years.

Imagine an apprentice learning this process for the first time in a workshop near Giza, four and a half thousand years ago. He would have spent hours pressing a hollow copper tube against a granite surface while a senior craftsman worked the wooden bow that rotated it, wet sand trickling constantly into the cutting groove. The noise was a low, continuous grinding rasp, the smell stone dust and water and the faint metallic sharpness of worn copper. Progress was measured in millimetres per hour, and the measure of a good day's work was a smooth, precisely cylindrical core of granite extracted without cracking or deviation from the intended line. The craftsman who trained him had learned the same way, and the one before that craftsman as well. Across generations of this kind of transmitted, embodied practice, the Egyptian stone-cutting tradition accumulated the standards that modern instruments can still measure in the surviving work.

For flat surface finishing, the experimental evidence suggests a combination of stone pounding tools, copper chisels used to remove larger irregularities, and extended lapping with abrasive slurries on flat reference surfaces to achieve the final precision. The lapping process, in which two surfaces are rubbed against each other with abrasive between them, is inherently self-correcting over time: high spots are abraded faster than low spots, and with sufficient patience the two mating surfaces converge on true flatness. This is the same principle used in modern optical surface grinding, and it requires no technology beyond an abrasive and a reference flat. What it requires in abundance is time and systematic patience. The Egyptian construction tradition, operating at the scale of multi-decade projects with organized workforces of thousands of skilled craftsmen, had both.

The transport of large stone elements from quarry to construction site is the aspect of Egyptian monumental building that most frequently attracts incredulous attention, because the numbers involved are genuinely impressive by any standard. The Great Pyramid contains approximately 2.3 million stone blocks, with an average weight estimated at around 2.5 tonnes, and the largest stones in the structure weigh considerably more, with the granite beams above the King's Chamber estimated at between forty and sixty tonnes each. Moving a sixty-tonne granite beam from Aswan to Giza, a distance of roughly eight hundred kilometres, using the technology available to Old Kingdom Egypt, is not a trivial engineering problem even with a very large workforce. Understanding how it was done requires understanding the specific conditions that made Egyptian large-scale transport possible, conditions that are frequently overlooked in popular discussions that focus on the weight of the stones while ignoring the infrastructure that moved them.

The Nile was the key. The Nile flood plain provided a reliable annual transport corridor from the quarries of the south to the construction sites of the north, and Egyptian engineers were expert at building and using large wooden barges capable of carrying enormous stone loads. The annual flood of the Nile raised water levels throughout the valley and allowed barges to be brought to within a short distance of both the quarry site at Aswan and the pyramid construction site at Giza, minimizing the overland transport distance that was by far the most labor-intensive part of the operation. Reliefs from the tomb of Djehutihotep at Deir el-Bersha, dating to the Middle Kingdom, show a large statue being transported overland on a sledge pulled by hundreds of workers, with a figure at the front of the sledge pouring liquid, almost certainly water or a lubricant of some kind, onto the surface ahead of the runners to reduce friction. This is not a mysterious technology. It is applied physics, and it is exactly what modern engineering analysis predicts would have been necessary to move large masses on sledge runners over prepared surfaces.

Experimental reconstructions of overland stone transport using sledges, rollers, and lubricated tracks have confirmed that organized teams of workers can move extremely large loads with these methods, provided the ground surface is adequately prepared, the lubricant is applied consistently, and the operation is well coordinated. The organizational challenge is the limiting factor, not the physics. Moving a sixty-tonne block requires not merely pulling power but the logistical capacity to coordinate the workers, supply the lubricant, maintain the sledge and the prepared road surface, and manage the dozens of variables that affect the operation in real time. The pyramid builders had the organizational capacity for exactly this, as demonstrated by the workers' village discovered near Giza in the 1990s, a permanent settlement with bakeries, breweries, a medical facility, and the infrastructure of a large organized workforce fed, housed, and managed by a state apparatus of considerable sophistication.

The alignment of the Great Pyramid is another aspect of the construction that merits serious attention, because it is exceptional by any standard. The four base sides of the pyramid are oriented to the cardinal directions with an accuracy of better than four arc minutes of deviation from true north, an alignment that requires both a reliable astronomical reference and a method for translating that reference into a ground survey over a base length of approximately 230 metres. The most plausible explanation for achieving this level of directional accuracy, using methods available in ancient Egypt, involves astronomical observation, most likely of stellar transits or of specific stars near the celestial pole, combined with careful bisection of the observed arc to establish true north. The geometry required to perform this operation correctly is not advanced by modern standards, but performing it in practice, in the open air, at night, with only simple instruments, to a tolerance of four arc minutes over a 230-metre baseline, requires considerable skill and procedural care.

The leveling of the base, which across the full 230-metre expanse of the pyramid's footprint deviates from true level by no more than 2.1 centimetres, is equally impressive. The most likely method involves water leveling: the use of flooded trenches or shallow pools of water to establish a reference level and then the careful adjustment of the prepared bedrock surface to that reference. Water seeks its own level with perfect consistency, and a system of connected trenches filled to a common water surface provides an accurate level reference over large distances. Translating that reference into a prepared stone surface to the precision demonstrated at Giza still requires skilled work and careful monitoring, but it does not require any technology beyond a supply of water, basic leveling instruments, and craftsmen who knew what they were doing and had done it before.

The pseudoscientific literature on Egyptian construction focuses heavily on what it frames as impossibilities: the weight of the stones, the precision of the joints, the accuracy of the alignment, the speed of construction if the commonly cited completion time of twenty years is accepted. Each of these is presented as evidence that conventional explanations are inadequate and that some other, typically exotic, agency must have been involved. This framing persistently misrepresents what mainstream archaeology actually claims. The mainstream position is not that Egyptian construction was easy or unremarkable. It is that Egyptian construction was extraordinarily difficult, accomplished through the systematic application of methods that are understandable in terms of available materials and tools, organized by a state with the administrative capacity to direct large workforces over long periods, and driven by a cultural and religious imperative that made the investment of enormous resources in monumental building not merely possible but central to the functioning of the society.

None of this diminishes the achievement. The Great Pyramid is one of the most ambitious engineering projects in human history, and the precision of its construction, examined honestly and in detail, is more impressive rather than less when understood in terms of what it actually required: tens of thousands of craftsmen working for decades under organized direction, applying skills that had been developed and refined over centuries of Egyptian construction tradition, to a standard of quality maintained consistently across millions of individual operations. The wonder is not that they needed help from outside the human world. The wonder is that human beings, organized by a sophisticated state and motivated by a religious tradition that placed monumental construction at the center of royal identity, were capable of this at all. That capability, understood properly, is more remarkable than any supernatural explanation, because it tells us something true and verifiable about what organized human effort, applied patiently and skillfully over time, can actually accomplish.

The Egyptian stone-cutting tradition did not vanish in a single catastrophic event. It declined gradually as the conditions that had supported it changed: the weakening of the centralized state that organized and financed monumental construction, the disruption of trade networks that supplied particular materials, the shift in religious and political priorities that had made pyramid building the central project of the Old Kingdom. The knowledge of how to work stone to a high standard was never entirely lost from Egypt; craftsmen continued to work stone throughout the later periods of Egyptian history and into the Roman and Byzantine eras. What was lost was the organizational capacity to deploy that knowledge at the scale of the pyramid-building tradition, and with it the accumulated institutional wisdom about how to manage a multi-decade monumental construction project involving hundreds of thousands of person-hours of skilled labor. The methods survived in the hands of craftsmen; the system that had mobilized those craftsmen at the scale required for the greatest monuments did not.

Seven

The Nazca Lines and the Engineering of the Sky

From the ground, most of them are invisible. Walking across the Nazca plateau in southern Peru, a visitor moves through a landscape of dark reddish pebbles scattered across pale yellowish-grey soil, the surface broken occasionally by shallow depressions or low embankments, the geometry of the lines only occasionally apparent as a slight straightness in the terrain that the eye registers without quite resolving. The plateau is dry, almost entirely without vegetation, swept by winds that have been shaping the desert for millennia without disturbing the surface markings in any fundamental way. There are no trees to provide elevation, no hills immediately adjacent to the figures that allow a clear view of the whole. The people who made these marks, who spent generations drawing shapes across a desert floor that only becomes fully legible from the air, were working without any possibility of seeing the complete result of their labour from above. That fact, once properly absorbed, reframes every question about how and why the Nazca lines were made.

The lines were created by a culture archaeologists call the Nasca, a coastal and highland civilization of southern Peru that flourished roughly between 100 BCE and 800 CE, a period of approximately nine hundred years during which successive generations maintained, extended, and elaborated the geoglyphs of the pampa. The Nasca were not a monolithic civilization with a single political structure; the archaeological evidence suggests a society organized around regional centres with shared cultural traditions, including the ceramic tradition that has allowed researchers to date the geoglyphs by the pottery found at their surface and edges. The lines were not created all at once. They accumulated across centuries, with different periods of activity producing different kinds of marks, from the earliest simple lines and geometric shapes to the more elaborate animal and human figures that are now the most photographed and discussed. The plateau itself was chosen deliberately for this purpose; its specific microclimate, essentially rainless, with relatively stable winds that redistribute rather than erode the surface pebbles, has preserved the markings with an integrity that would have been impossible in almost any other environment on earth.

The technique by which the lines were created is well understood, and it is considerably less mysterious than popular accounts often suggest. The surface of the pampa is covered by a layer of dark iron-oxide-coated pebbles that sit atop a much lighter substrate of yellowish clay and calcium carbonate. The Nasca lines were made by removing the dark surface pebbles and piling them in low banks along the edges of the cleared areas, exposing the light substrate beneath. The resulting contrast, dark borders framing a pale interior, is what makes the lines visible from the air and, for the broader cleared areas, from ground level at some distance. The tools required for this process were minimal: baskets to carry the removed pebbles, wooden or bone rakes to sweep the surface, and the organizational capacity to keep a large number of workers moving in coordinated patterns across the plateau. No metallurgy, no sophisticated machinery, no elevated vantage point was necessary for the basic clearing operation.

What was necessary, and what represents the genuine engineering achievement of the Nazca lines, was the surveying and planning system that allowed the Nasca to project geometrically precise lines across distances of several kilometres, to maintain consistent angles over terrain that is not perfectly flat, and to construct the large animal figures with proportions that remain recognizable and coherent when viewed from high altitude. A straight line several kilometres long, drawn across a surface with natural irregularities, does not stay straight without a consistent method of alignment.

A spider figure whose legs and body must correspond to a coherent animal form across a cleared area hundreds of metres wide cannot be laid out accurately without some system for translating a small-scale design onto a large-scale surface. These problems are real, and their solutions reveal a level of mathematical and geometric knowledge that the Nasca possessed and applied with considerable skill.

The most plausible explanation for the long straight lines, supported by experimental work by researchers including Joe Nickell and the team assembled by NOVA for a practical reconstruction project in the 1980s, involves the use of wooden stakes and cords stretched between them to establish direction and then extended by sighting along the established line. A surveying team with a baseline cord, a series of intermediate stakes, and the discipline to keep each new stake on the sight line of the previous ones can project a straight line across several kilometres with sufficient accuracy to match what is observed on the plateau. The technique requires no instruments beyond cord, stakes, and careful human attention, and it is the kind of surveying method that would have been independently available to any culture with the organizational capacity and the motivation to use it systematically over a large area.

The construction of the large figurative geoglyphs, the hummingbird, the spider, the condor, the monkey, the whale, required a somewhat more sophisticated approach. Experimental archaeologist Joe Nickell demonstrated in the 1980s that the figures could be laid out by scaling up a small design using a grid system, transferring the proportions of each grid square onto the ground using measured cord lengths and stakes. The Nasca had access to the mathematical concepts required for this: proportional scaling is a geometric operation that follows directly from the understanding of similar triangles, and the Nasca ceramic tradition demonstrates a culture with a highly developed sense of proportion and pattern applied consistently at different scales. The monkey figure, which is among the largest and most complex of the animal geoglyphs, has been analyzed by several researchers and found to be geometrically consistent with a grid-based enlargement of a design that could have been sketched on a much smaller surface before transfer to the plateau.

A surveying party beginning work on one of the larger figures would have started with a reference design, probably drawn on cloth or ceramic, and a measured baseline established on the ground. The team leader would have calculated the scaling factor and divided the design into a grid of proportional squares, then directed workers to clear the surface within each square's corresponding ground area according to the lines in the design. The work would have been slow, methodical, and highly organized, requiring clear communication between the person holding the reference design and the teams clearing the surface, and constant checking of measurements to prevent cumulative errors from distorting the figure. It was demanding work, but it was a known kind of demanding: the kind that a culture with a tradition of large-scale organized labour, sophisticated ceramic design, and generations of experience working on the plateau could accomplish without any technology that we would not expect them to have had.

The question of why the Nasca made the lines has generated considerably more scholarly debate than the question of how, and it remains less settled. Several interpretive frameworks have been proposed, each with genuine supporting evidence and each with limitations that keep the question open. The astronomical interpretation, associated particularly with Paul Kosok and Maria Reiche, who devoted much of her life to studying and protecting the lines, proposed that the geoglyphs served as a kind of astronomical calendar, with specific lines aligned to the rising and setting points of important celestial bodies at significant times of the year. Some alignments do exist between particular lines and astronomical events, but statistical analysis by the archaeoastronomer Anthony Aveni and his colleagues found that the number of astronomically significant alignments is not substantially greater than would be expected by chance given the large number of lines and the many possible celestial reference points. This does not rule out astronomical significance for specific individual lines, but it undermines the claim that the whole system was primarily an astronomical device.

The water and irrigation hypothesis, developed most fully by the archaeologist David Johnstone and others working on the relationship between the Nasca lines and the underground water channels called puquios that supplied the coastal desert settlements, proposes that the lines mapped or marked water sources, aquifer locations, or paths of ritual procession connected to water worship. The pampa is one of the driest places on earth, and water was the fundamental resource around which Nasca settlement and agriculture were organized. The underground aqueduct system that the Nasca constructed to bring water from the Andes foothills to the coastal plain is itself an extraordinary engineering achievement, and it is entirely plausible that a culture so deeply invested in water management would have incorporated water-related symbolism and ritual into a landscape as prominent and deliberately marked as the plateau. The location of some geoglyphs near or above underground water features provides some supporting evidence, though the correlation is not consistent enough to serve as a complete explanation.

The most broadly supported current interpretation treats the Nazca lines as primarily ritual spaces, surfaces for ceremonial activity that combined cosmological, agricultural, and social functions in ways that resist reduction to any single explanatory category. Ceramic and textile fragments found on the surface of the geoglyphs suggest that they were used for ritual walking and possibly for offerings, and the trapezoid shapes in particular, the widest and most cleared of the geometric figures, may have served as assembly areas for ceremonial gatherings. The animal figures may represent totemic or cosmological symbols connected to the religious traditions of the Nasca world, where the condor, the hummingbird, the spider, and the monkey all carried specific cultural meanings documented in the ceramic and textile record. From this perspective, the lines were not a message to extraterrestrial visitors, not a landing strip for aircraft, and not primarily a scientific instrument, but a sacred landscape, a deliberately constructed environment in which ritual activity took place and through which the Nasca maintained their relationships with the natural and supernatural forces that governed their world.

The extraterrestrial hypothesis, which attributes the lines to alien visitors because of the assumption that they could only be appreciated from the air, rests on a misunderstanding that the preceding analysis should make clear. The Nasca did not need to see the finished figures from the air in order to make them. They needed to know what they were making, which could be encoded in a small-scale design, and they needed the surveying and organizational skills to transfer that design accurately to the ground, which they demonstrably possessed. The idea that ground-level humans cannot appreciate large-scale landscape designs is also simply false: indigenous cultures around the world have created large earthworks, stone arrangements, and landscape modifications whose meaning was fully understood by the people who made and used them without any aerial perspective. The lines were made by people, for purposes that made complete sense within their cultural and religious world, using techniques that are entirely explicable in terms of the knowledge and resources available to them. That explanation is less exciting to a certain kind of imagination than the alien alternative, but it has the substantial advantage of being true.

The Nasca plateau today is a UNESCO World Heritage Site, protected from the construction and agricultural activity that threatened it through much of the twentieth century. Maria Reiche spent decades living near the lines, studying them, and advocating for their preservation, and her work established the foundation on which subsequent archaeological and scientific research has built. Modern remote sensing, including satellite imaging and aerial photogrammetry, has revealed new geoglyphs previously unknown and allowed the existing figures to be mapped with a precision that earlier researchers could not achieve. The picture that has emerged is of a landscape that was continuously modified over nine centuries by a culture that treated the plateau as a living, accumulating record of their cosmological and social world, adding new figures and lines in each generation while maintaining and re-walking those created by their ancestors. Understanding the Nazca lines in their full complexity requires holding all of this together: the geometry, the labour, the ritual purpose, the water preoccupation, the cosmological symbolism, and the nine centuries of accumulated meaning that make them, in the end, one of the most sustained acts of landscape inscription in human history.

What the lines reveal about the Nasca as engineers and planners is not that they possessed some mysterious elevated capability unavailable to ordinary human cognition. It is that they possessed, in abundance, the qualities that distinguish all the technical traditions examined in this book: patience, organizational discipline, accumulated knowledge applied systematically over long periods, and a cultural commitment to a large-scale project that outlasted any individual lifetime by many generations. The lines were not built by a single inspired genius or a single extraordinary generation. They were built by a culture that maintained a practice across nine centuries, adapting and extending what had been done before, and leaving in the desert a record of their priorities and their cosmological imagination that the desert has preserved, against all probability, until the present day.

Eight
Ancient Automata and Mechanical Life

Sometime in the first century CE, in the city of Alexandria, an engineer and mathematician named Hero assembled a small wheeled cart, loaded it with a coiled rope connected to a falling lead weight, and ran it across the floor of his workshop. The cart moved forward, turned at a predetermined angle, moved again, turned again, and traced a path across the floor that Hero had designed in advance by adjusting the rope configuration before the device was set in motion. It was, by any reasonable definition, a programmable vehicle: a machine whose sequence of movements had been encoded in its mechanical setup before the operation began and that executed those movements without any further human intervention. Hero described it in his treatise on mechanics, now called the Automata, with the matter-of-fact precision of a man reporting a solved engineering problem rather than announcing a wonder. For him, the programmable cart was one demonstration among many of the principles he was interested in, a worked example in a larger project of understanding what mechanical systems could be made to do.

Hero of Alexandria is the figure most closely associated with ancient automata in the modern scholarly literature, and his surviving works represent a remarkable body of technical writing that covers pneumatics, mechanics, optics, surveying instruments, and theatrical machinery with a thoroughness that has no parallel in the ancient world outside of Archimedes. But Hero was not working in isolation, and the tradition of mechanical devices he contributed to was considerably older than the first century CE. The roots of ancient automata reach back at least to the fourth century BCE, and possibly further, into the technological culture of classical and Hellenistic Greece, where the boundary between engineering, philosophy, natural inquiry, and religious practice was considerably more permeable than it would later become in the modern West.

The philosopher Archytas of Tarentum, a contemporary and friend of Plato who worked in the early fourth century BCE, is credited in ancient sources with constructing a mechanical dove capable of flight. The device is described as being driven by compressed air or steam, suspended on a balanced beam or pivot, and capable of flapping its wings in a simulation of avian movement. The ancient sources that mention it, including Aulus Gellius and Favorinus writing several centuries after the fact, treat it as a genuine historical achievement rather than a myth, and while the technical details are too sparse to allow confident reconstruction, the general principle, a bird-shaped mechanism driven by a pneumatic or steam-based power source, is entirely consistent with the engineering knowledge available in fourth-century BCE Greece. Archytas was deeply interested in the relationship between mathematical principles and mechanical applications, and the dove, whether or not it achieved anything like sustained free flight, represents an early attempt to encode the appearance of biological motion in a mechanical device.

The philosophical dimension of this project is important for understanding why ancient automata were made and what their makers thought they were doing. In the Greek intellectual tradition, particularly in the Aristotelian framework that dominated natural philosophy from the fourth century BCE onward, the distinction between natural and artificial was not a categorical boundary but a question of degree and origin. Aristotle himself observed that art, meaning skilled human making, imitates nature, and he applied this principle to mechanical devices as well as to sculpture and architecture. A mechanical bird that flapped its wings was, from this perspective, not a violation of the natural order but an imitation of it, a demonstration that the principles governing biological motion could be understood well enough to reproduce some of their effects through artificial means. The attempt to create mechanical life was, in this framework, simultaneously a philosophical investigation into what life is and an engineering demonstration of how far human skill could approximate it.

Hero's Automata presents two major categories of self-moving devices. The first are theatrical machines: complex mechanical setups designed to produce dramatic performances for audiences at religious festivals or theatrical events. Hero describes in detail a self-contained theatrical display in which a miniature stage, driven by a falling weight and a system of strings and pulleys, performed a multi-scene play involving moving figures, sound effects produced by mechanical means, and lighting changes achieved through the controlled burning of materials in small enclosed chambers. The figures on the stage were moved by cords attached to the central mechanism, their gestures and movements coordinated with the narrative of the play through the precise adjustment of cord lengths and attachment points before the performance began. Once started, the device ran the entire performance without human intervention, the falling weight gradually unwinding and driving the sequence of mechanical events that Hero had programmed into it.

The second category in Hero's Automata consists of self-moving vehicles, including the programmable cart described at the opening of this chapter. Hero explains the mechanical principle clearly: a rope is wound around an axle in a specific pattern, and as a lead weight causes the rope to unwind, the axle rotates and drives the wheels. By varying the direction of the rope's winding, the designer can specify whether the cart will move forward or turn, and by adjusting how many turns of rope are wound in each configuration, the distance traveled in each direction can be controlled. The device is, in modern terms, an open-loop control system: the program is set before operation begins, there is no feedback mechanism to correct errors during execution, and the accuracy of the performance depends entirely on the precision with which the mechanical setup encodes the intended sequence. Hero understood the concept well enough to work out the mathematics of the rope-winding system in detail, specifying how many turns of the axle correspond to a given distance of travel and how the turning radius depends on the differential rotation of the two drive wheels.

The pneumatic devices described in Hero's other major work, the Pneumatica, extend the range of ancient mechanical ingenuity into a different domain. The Pneumatica catalogs more than a hundred devices that use compressed air, steam, siphon action, or vacuum effects to produce movement, sound, and apparent miracles in temple and theatrical settings. Among them are coin-operated holy water dispensers, in which dropping a coin into a slot causes a measured amount of liquid to be released by a counterweighted valve; singing mechanical birds driven by water displacement through air channels; temple doors that open automatically when a fire is lit on an altar, with the heat expanding air in a sealed chamber below the altar and forcing water into a counterweighted bucket that pulls the door open; and a device for producing thunder sounds during theatrical performances by releasing a cascade of bronze balls onto a stretched membrane. These are not toys or curiosities in Hero's presentation. They are engineering demonstrations, worked examples of the principles of pneumatics and hydraulics applied to the specific demands of religious spectacle and theatrical performance.

The temple door device deserves particular attention because it illustrates a dimension of ancient automata that is easy to overlook when focusing on the mechanical ingenuity alone: the social and religious context in which these devices functioned. A worshipper approaching a temple whose doors appeared to open of their own accord when the offering fire was lit was experiencing what the priests and engineers who managed the temple intended as a demonstration of divine presence or divine approval of the sacrifice. The mechanical explanation was known to the temple authorities and presumably to Hero, who describes the device technically without any apparent attempt at mystification. But for the worshippers who experienced the effect without knowing its mechanism, the opening doors were a genuine religious experience, a confirmation that their offering had been received and that the god was present. Ancient automata were not merely engineering achievements; in their most important deployments they were instruments of religious authority, using mechanical knowledge to manage the experience of the divine.

Hero imagines his audience as engineers and educated readers interested in mechanical principles rather than as credulous worshippers to be deceived, and he presents the temple devices with the same dispassionate technical clarity he applies to the theatrical machines and the programmable vehicles. This tone of matter-of-fact mechanical exposition is itself revealing. It suggests that by the first century CE, within the engineering culture of Hellenistic Alexandria, mechanical devices of considerable sophistication were regarded as a normal part of the intellectual and practical landscape, not as marvels requiring special justification but as applications of understood principles to the specific demands of different contexts. The automaton tradition that Hero was documenting and extending had been developing for at least four hundred years before his time, and the accumulated knowledge it represented was sufficiently established that he could treat it as a foundation to build on rather than as a novelty to explain.

The wider ancient world outside Alexandria also produced mechanical devices of notable sophistication, though the surviving record is patchier and the technical descriptions less precise. The ancient Chinese tradition of mechanical engineering produced water-driven astronomical clocks and mechanical figures as early as the Han dynasty, with the engineer Zhang Heng constructing a water-powered armillary sphere in the second century CE that rotated in sync with the actual celestial sphere above it. The Islamic engineering tradition, building on Greek and Persian foundations in the medieval period, produced a remarkable body of work culminating in al-Jazari's Book of Knowledge of Ingenious Mechanical Devices, written in 1206 CE, which describes fifty mechanical devices including water clocks, hand-washing automata, musical boats, and a programmable drummer that could be adjusted to play different rhythmic patterns by rearranging pegs on a rotating drum, an early example of what modern engineers would recognize as a programmable controller.

The relationship between the ancient Greek automaton tradition and these later developments is partly one of direct transmission and partly one of parallel development from shared principles.

The Islamic engineers of the medieval period had access to Arabic translations of Hero's works, and the pneumatic and hydraulic principles he described appear in various forms throughout the medieval engineering literature. The Chinese tradition developed more independently, though there is evidence of some contact between Chinese and Hellenistic engineering cultures along the Silk Road. What is clear is that the desire to create self-moving, life-imitating machines was not confined to any single civilization or historical period but appeared repeatedly wherever cultures developed sufficient mathematical and mechanical knowledge to make the attempt practicable.

What was lost from the ancient automaton tradition is not the basic principles, which survived in various forms through the medieval period and were fully recovered and extended in the Renaissance and the early modern era. What was lost was the specific accumulated knowledge of particular devices, the workshop traditions that maintained and improved them, and the institutional contexts that supported their production. Hero's Pneumatica describes devices that were clearly the products of an active engineering culture with the tools, materials, and organizational capacity to build complex mechanisms. When the economic and institutional conditions of Hellenistic Alexandria changed, when the network of workshops and patrons that supported this kind of engineering work was disrupted, the specific devices disappeared along with the culture that had produced and used them. The principles were recoverable from texts; the craft knowledge that allowed skilled builders to translate those principles into functioning objects was not.

A craftsman working in Hero's workshop would have spent years learning to cut and fit the small brass components that his pneumatic devices required, to seal joints with the precision needed to maintain air pressure across a complex valve system, to calibrate the weight and movement of counterbalances to produce the timing effects his devices depended on. Sitting at his bench in the late afternoon light filtering through a high window, testing a valve assembly for the fourth time while adjusting the seating with a fine file, he was applying knowledge that could only be built through practice, the accumulated sensitivity of hands that had made many such adjustments and learned, through failure and correction, exactly how much was enough. That knowledge was not in Hero's texts. It was in the workshop, in the hands of the people who worked there, and it dispersed along with them when the conditions that had made their work possible changed beyond recovery.

The automata tradition is, in this respect, a particularly clear illustration of the gap between the knowledge that survives in texts and the knowledge that lives in practice. Hero's descriptions are precise enough to allow modern engineers to understand the operating principles of his devices and to construct working replicas, as several research groups have done. But those replicas are achievements of modern engineering knowledge applied to ancient designs, not continuations of the ancient craft tradition. The hands that made them had learned their skills in entirely different contexts and brought entirely different accumulated experience to the work. What the ancient craftsmen who worked with Hero knew, and what their successors in the workshops of late antiquity knew, about the specific materials, tolerances, and adjustments required to make these devices work reliably and well, belongs to the long list of technical knowledge that history has not preserved.

Nine
Hydraulic Wonders of the Ancient Cities

On a summer morning in the second century CE, a resident of Rome woke to running water. Not the collected rainwater or the carried amphora that remained the daily reality for most of the ancient world, but water that arrived under pressure through a lead pipe connected to a distribution system that drew from a source in the hills outside the city, moved through a network of stone-lined channels sometimes cut through solid rock, and carried across valleys on multi-tiered arched aqueducts to street-level fountains, public baths, and, for the wealthiest citizens, directly into private homes.

The supply was essentially continuous. The pressure, in the better-served parts of the city, was sufficient to deliver water to upper floors. The volume delivered daily to Rome at its height has been estimated by the historian of technology A. Trevor Hodge at approximately one million cubic metres per day, enough to supply more than a million people with a per capita daily supply comparable to that of a modern European city. This was not an accidental achievement. It was the product of several centuries of systematic hydraulic engineering, informed by empirical observation, mathematical calculation, and an organizational capacity that made the construction and maintenance of the system possible as a continuous civic enterprise.

The Roman aqueduct system is the most famous expression of ancient hydraulic engineering, but it was neither the first nor the only one. Ancient water management had deep roots in every region where large populations concentrated in arid or semi-arid environments, and the specific technical solutions developed in different civilizations, while varying considerably in their details, share a common foundation of careful observation of how water moves, how it can be stored and directed, and what structures will reliably contain and channel it over long periods of use. The Mesopotamian irrigation systems of the third millennium BCE, which transformed the alluvial plains of the Tigris and Euphrates into one of the most productive agricultural regions in the ancient world, represented hydraulic engineering on a scale and of a complexity that required organized administrative oversight, systematic maintenance, and a technical knowledge tradition capable of designing and correcting large irrigation networks. The Minoan palace at Knossos on Crete, occupied from roughly 1900 to 1450 BCE, had a drainage and water supply system whose complexity astonished the archaeologists who excavated it in the early twentieth century, including terracotta pipes with parabolic cross-sections designed to maintain water velocity and prevent sediment accumulation.

The qanat, the underground channel technology developed in ancient Iran and subsequently spread across the arid regions of the Middle East, Central Asia, North Africa, and the Iberian peninsula, represents one of the most sophisticated and widely distributed hydraulic technologies of the ancient world. A qanat is an underground gallery, typically one to two metres high and wide enough for a person to work in, cut through the alluvial fans of mountain piedmont areas to tap the shallow aquifers that accumulate where highland streams lose their velocity and deposit their sediment loads.

The gallery descends at a slight gradient less steep than the land surface, allowing water to flow by gravity from the aquifer to an outlet at the surface some distance away, sometimes several kilometres from the collection point. The vertical shafts sunk at regular intervals to allow construction access and ventilation are a characteristic feature of the qanat landscape, visible from the air as dotted lines of circular depressions running across the terrain. The oldest reliably dated qanats in Iran are approximately three thousand years old, and at the height of their distribution there were tens of thousands of them across the ancient Near East and beyond, delivering water to cities, farms, and caravanserais across some of the most inhospitable terrain in the world.

The engineering demands of qanat construction were considerable. The gradient of the gallery had to be controlled with considerable precision: too steep and the water velocity would erode the floor; too shallow and the water would stagnate and the gallery would fill with sediment. The survey required to establish and maintain the correct gradient over several kilometres of underground excavation, using only plumb lines, spirit levels, and measurements from the surface shafts, represents a sophisticated practical geometry applied in the dark, often in cramped and poorly ventilated conditions. The workers who dug qanats, the muqannis, formed a specialized hereditary profession whose technical knowledge was transmitted entirely through apprenticeship and practice. Their understanding of aquifer behaviour, of the geological conditions that indicated productive water-bearing strata, of the engineering requirements for stable gallery construction in different soil types, and of the maintenance procedures necessary to keep an operational qanat clear of sediment and structurally sound was among the most sophisticated practical hydrogeology in the pre-modern world. Much of it was never written down.

Roman hydraulic engineering built on earlier traditions while developing distinctive organizational and technical capabilities of its own. The eleven major aqueducts that served Rome by the late imperial period ranged in length from sixteen to ninety-one kilometres, and their construction involved challenges of surveying, materials engineering, and logistics that were formidable even by the standards of modern infrastructure projects. The surveying problem alone was substantial: an aqueduct must descend at a consistent, precisely controlled gradient from its mountain source to its urban destination, deviating neither too steeply, which would cause damaging water velocities, nor too shallowly, which would cause the water to pool and stagnate. Frontinus, the Roman water commissioner who wrote a detailed account of the Roman water supply system around 97 CE, records that his predecessors had sometimes made surveying errors that required expensive corrections, and the physical evidence from surviving aqueduct channels confirms that gradient control was an ongoing challenge that the Roman engineers addressed with varying success across different structures.

The Roman solution to the problem of carrying an aqueduct channel across a valley that interrupted the natural slope of the terrain was the arcade, the multi-tiered arched bridge structure that has become the most recognizable symbol of Roman engineering. The Pont du Gard in southern France, built in the first century CE to carry the Nimes aqueduct across the Gardon river valley, rises to a height of nearly fifty metres in three tiers of arches and has been standing for nearly two thousand years without significant structural modification. It is not the longest or the tallest Roman aqueduct arcade, but it is among the best preserved and most studied, and its engineering has attracted sustained attention from structural engineers who find in it a near-perfect expression of Roman arch construction principles. The voussoir arch, in which wedge-shaped stone elements transfer loads laterally to the supporting piers, is an inherently stable structure when correctly proportioned, and the Roman engineers who built the Pont du Gard understood the proportions required for stability with a precision that the structure's two-thousand-year survival confirms empirically.

At the city end of the distribution system, the water entered a castellum, a distribution tank in which it was divided among different categories of use and directed through different pipe networks to different parts of the city. The castellum divided the supply into three channels: one for the public fountains that were the primary water source for most of the urban population, one for the public baths that were a central institution of Roman civic life, and one for private connections to the houses of wealthy citizens. This three-tier distribution system encoded a social hierarchy of access into the physical infrastructure of the water supply: public fountains first, baths second, private connections third, with the understanding that if the supply was insufficient to serve all three, the private connections were cut off before the public fountains and the baths. Frontinus describes this priority system explicitly, and it reflects a hydraulic management philosophy that treated the reliable supply of public infrastructure as a higher obligation than the comfort of private consumers.

The pipes that carried water within the city were made primarily of lead, cast in sheets and soldered into cylindrical pipes of various standard diameters. Lead pipe was expensive, heavy, and required skilled workers to produce and install, but it was chemically stable in contact with Roman water chemistry, flexible enough to be bent around obstacles, and joinable with solder in ways that allowed the distribution network to be extended and modified without replacing entire sections. The concern that Roman lead pipes may have contributed to lead poisoning in the population is a recurring topic in discussions of Roman public health, and while lead dissolved from the pipe walls was certainly present in Roman tap water, recent geochemical analysis of Roman water deposits suggests that the calcium carbonate hardness of the water in most Roman systems would have coated the inside of the pipes relatively quickly, forming a protective layer that significantly reduced lead dissolution after the initial period of use.

The public baths, the thermae, were the most technically demanding endpoint of the Roman water supply system, requiring not only reliable delivery of large water volumes but a complex infrastructure for heating the water and maintaining the different temperature zones that characterized the Roman bathing sequence. The Baths of Caracalla in Rome, completed in 217 CE, covered an area of approximately thirteen hectares and could accommodate sixteen hundred bathers simultaneously in a sequence of spaces moving from the cold room through the warm rooms to the hot room, with outdoor exercise areas, libraries, shops, and social spaces incorporated into the complex. The hypocaust heating system, which circulated hot gases from a furnace through a raised floor and hollow wall systems throughout the heated rooms, was itself a sophisticated engineering achievement, but the hydraulic demands of the baths were equally impressive: the continuous supply of cold water for the cold pool, the hot water for the caldarium, the warm water for intermediate spaces, and the drainage of the used water into the city's sewer system all required precise hydraulic calibration maintained continuously by a staff of engineers and slaves who understood the system well enough to keep it running.

Below the streets of Rome, the Cloaca Maxima, the great drain that had been carrying waste water and storm runoff from the city center to the Tiber since the sixth century BCE, formed the backbone of a drainage system that was extended, deepened, and elaborated across the centuries of the republic and empire. The Cloaca Maxima in its final form was a barrel-vaulted tunnel of sufficient diameter that a small boat could be rowed through it, and its engineering was robust enough that substantial sections of the ancient structure are still in service in the modern drainage system of Rome. Pliny the Elder, who marveled at many Roman engineering achievements, considered the Cloaca Maxima among the greatest, noting that it had been built to handle the drainage of seven hills and that it had performed this function without significant failure for more than six hundred years before his own time.

The hydraulic traditions of other ancient civilizations are less well documented than the Roman system but no less impressive in their own contexts. The Nabataean civilization of the Arabian desert, centered on the rock-cut city of Petra in what is now Jordan, developed one of the most sophisticated water harvesting and distribution systems in the ancient world, channeling flash flood water from a wide catchment area through carved rock channels, storing it in rock-cut cisterns, and distributing it through ceramic pipe networks to the various districts of the city. The Nabataeans occupied a region with annual rainfall that would be considered severe drought conditions almost anywhere else, and their survival as a prosperous trading city in the middle of the desert depended entirely on their ability to capture and store every drop of water that fell in their catchment area. The channels, sluices, and cisterns they built to accomplish this are carved directly into the sandstone of the surrounding landscape with a precision and completeness that reflects generations of accumulated hydraulic knowledge specifically adapted to their extreme environment.

The Sri Lankan civilization of the early common era constructed a system of large artificial reservoirs, called tanks, connected by channels and sluices across a landscape in which the timing and distribution of rainfall was highly variable, storing monsoon rainwater for use in the dry season and distributing it through gravity-fed irrigation canals to paddy fields that supported dense agricultural populations. The tanks of the Anuradhapura period, some of which covered areas of more than two thousand hectares, were maintained by a system of sluices incorporating a device called the bisokotuwa, a valve chamber built into the base of the dam that allowed flow rates to be controlled with considerable precision and that prevented the erosion of the dam structure by controlling the velocity of outflow water. The bisokotuwa is a genuine hydraulic engineering innovation, and its widespread application across the Sri Lankan tank system represents organized technical knowledge applied systematically to a specific hydraulic management problem.

What is striking about ancient hydraulic engineering, when surveyed across these different civilizations, is how consistently the same basic engineering intuitions appear in different forms adapted to radically different environments. The understanding that water flows downhill at a rate related to its gradient, that channels must be designed to maintain the right velocity to transport water without eroding their beds or depositing sediment, that storage structures must be built with control mechanisms to manage outflow, that distribution networks must prioritize uses in order to ensure reliable supply to the most important endpoints, all of these principles appear in Roman aqueducts, Iranian qanats, Nabataean cisterns, and Sri Lankan tanks. None of these cultures had formal hydrological theory in the modern sense. All of them had generations of empirical observation, practical experiment, and accumulated craft knowledge that converged on solutions to the same fundamental problems from different starting points and with different available materials.

The decline of ancient hydraulic infrastructure followed the decline of the political and economic systems that maintained it, and in most cases the deterioration was gradual rather than sudden. Roman aqueducts were not destroyed in a single catastrophic event; they were damaged by warfare and earthquake, fell into disrepair as the tax revenues that funded their maintenance evaporated, and were gradually abandoned as the population they served declined and the administrative capacity that had kept them running was redirected or extinguished. The Aqua Claudia, one of Rome's largest aqueducts, was already in significant disrepair by the fifth century CE, and Procopius records that during the Gothic wars of the sixth century the invading forces cut the aqueducts supplying Rome as a military tactic, reducing the city's water supply to the river and whatever cisterns had been maintained. The infrastructure could in principle have been repaired; what had been lost was the organizational and economic capacity to do so, the same kind of systemic loss that explains the disappearance of Roman concrete, Damascus steel, and so many other ancient technical achievements.

The water systems of the ancient world mattered not just as engineering achievements but as the fundamental infrastructure on which dense urban life depended. Without reliable water supply, the Roman city at its scale and density could not have existed. Without the qanat, the cities of the Iranian plateau and the oasis settlements of Central Asia could not have survived in their environments. Without the tank system, the paddy agriculture of Sri Lanka could not have supported its population. Water management was not a luxury or an enhancement of ancient urban life; it was the precondition for it, and the sophistication of the hydraulic engineering traditions that provided it reflects the depth of the problem and the intelligence and organizational capacity of the cultures that solved it. That those solutions were, in many cases, lost along with the political and economic systems that sustained them is one of the clearest demonstrations in the historical record of the principle that technical achievement is always embedded in a social context, and that when the context fails, the achievement does not automatically survive it.

Ten

Impossible Stones and Monumental Architecture

At Baalbek in the Bekaa Valley of modern Lebanon, three of the largest stones ever used in human construction sit in the western retaining wall of the Roman temple complex. They are known as the Trilithon, and they have been sitting in that wall for roughly two thousand years. Each one is approximately nineteen to twenty metres long, four metres high, and three and a half metres deep. Their individual weights have been estimated, on the basis of the specific gravity of the local limestone, at between eight hundred and a thousand tonnes each. They were quarried from a site roughly one kilometre from the temple platform, moved to the construction site, lifted to a height of approximately seven metres, and fitted into the wall with a precision that leaves joints of only a few millimetres between adjacent stones. A fourth stone, even larger, was partly quarried and then abandoned at the original quarry site; it lies there still, still attached to the bedrock at one end, and it is known as the Stone of the Pregnant Woman. It weighs approximately twelve hundred tonnes. Nothing built by the modern construction industry has moved a single stone of comparable weight.

The Trilithon is the most dramatic example of a phenomenon that appears, in various forms, across the megalithic and monumental building traditions of the ancient world: stones of enormous size, moved from quarry to building site, lifted, fitted, and incorporated into structures with a precision that seems disproportionate to the tools available and, in some cases, with a logistical complexity that challenges straightforward explanation even when the engineering principles are well understood. The word impossible, which appears constantly in popular writing about these structures, is almost never technically accurate. What it usually means is that the engineering required was very demanding, that the organizational scale was very large, and that the modern imagination, conditioned by machinery and motor transport, struggles to picture how the work was done by human muscle and lever and rope. Those are genuine challenges to understanding, but they are historical and imaginative challenges rather than physical impossibilities, and the distinction matters enormously.

Baalbek itself illustrates the challenge well. The Roman builders who incorporated the Trilithon into the temple platform were working in a tradition of organized large-scale construction that had been developed over centuries across the Roman world, and they had access to the full range of Roman engineering knowledge: organized labour, lifting machines based on pulley and capstan systems, wooden sledges, rollers, prepared road surfaces, and the institutional capacity to coordinate operations involving hundreds of workers over extended periods. The specific problem of moving stones weighing eight hundred to a thousand tonnes, however, pushed this tradition to a scale for which no directly comparable Roman example survives, and the methods used at Baalbek remain a matter of active scholarly investigation and some genuine uncertainty. What can be said with confidence is that the Romans solved the problem, that the Trilithon stones are where they are, and that the solution, whatever its details, fell within the range of what Roman engineering, applied at its maximum scale and organizational capacity, could achieve.

The polygonal masonry tradition of the ancient Mediterranean and Andean worlds presents a different kind of challenge. In structures ranging from the Mycenaean citadel walls of Bronze Age Greece to the Inca temples and enclosures of Cusco in Peru, large stones were shaped into irregular polygonal forms and fitted together without mortar in configurations of extraordinary interlocking complexity. In the finest examples, individual stones have six, eight, or more faces, each one meeting the faces of adjacent stones with joints so tight that a blade of grass cannot be inserted between them.

The geometry of the interlocking is three-dimensional, with stones shaped to engage each other in ways that resist both horizontal and vertical displacement. This is not masonry that can be achieved by cutting stones to approximate size and shimming with smaller pieces; it requires either that the stones be cut precisely to predetermined dimensions or that each stone be shaped individually to fit its specific location in the wall. The second approach, which appears to be what the best ancient practitioners used, requires a system for measuring and transferring the precise geometry of each contact face to the stone being shaped to fill it.

The Inca masonry tradition, which produced the finest surviving examples of polygonal stone fitting in the world, was carried out by craftsmen called pirca, working primarily in andesite, a moderately hard volcanic rock considerably less tractable than limestone but easier to work than granite. The Inca worked their stone without metal tools; their primary shaping implements were stone hammers of harder material, used as both percussion and abrasive tools to knock off high spots and grind faces to the required profile.

A pirca working on a stone destined for a specific place in a wall would have had, as his primary guide, the negative impression of the space the stone needed to fill: the faces of the stones already in place on either side and below the gap, which defined the geometry his stone needed to match. The fitting process was iterative, bringing the new stone into the approximate position, marking where it made contact and where gaps remained, removing it, grinding down the high spots, and bringing it back to check the fit again. The cycle was repeated, patiently and as many times as necessary, until the stone settled into contact with all its neighbours along every face simultaneously.

This process, which requires no technology beyond hard-stone hammers and the patience to repeat a fitting cycle many times, is entirely capable of producing the joint quality observed in the best Inca walls, given sufficient time and sufficiently skilled craftsmen. Experimental stonemasons have demonstrated this in practice: the fitting of polygonal stones to the Inca standard is slow work that demands attention and patience but no mysterious capability. The astonishment that visitors feel before a wall of perfectly fitted Inca masonry is an appropriate response to an extraordinary achievement, but the achievement is one of organized skill and accumulated craft tradition, not of supernatural assistance or lost technology in any meaningful sense. What the Inca had that made it possible was not a secret method but a professional tradition of stone-fitting craftsmen trained over generations, a state system that could deploy those craftsmen at scale and supply them continuously with materials and food, and a cultural commitment to quality in stone construction that made the investment of the required time and labour not just acceptable but obligatory.

The transport of large stones in the Andean context presents questions that remain more genuinely open than the fitting question. The largest stones in the Inca world are at Sacsayhuaman, the enormous terraced fortress above Cusco, where individual limestone blocks weigh up to three hundred tonnes and were moved from quarries several kilometres away across terrain that includes both flat ground and steep inclines. The Inca had no wheeled vehicles and no large draught animals comparable to the oxen used in Old World construction; their primary large animal was the llama, which can carry loads of perhaps forty kilograms and cannot be used for hauling heavy stones. They did have ropes, wooden rollers and sledges, large organized workforces, and the institution of mit'a, the labour tribute required of subject peoples that provided the Inca state with access to very large numbers of workers for specific projects. The combination of these resources, applied to the problem of moving three-hundred-tonne blocks up steep Andean slopes, implies either a ramp and sledge system of considerable engineering sophistication or some other mechanical arrangement that has not been identified with certainty from the archaeological evidence.

Experimental reconstructions of Inca stone transport using rope, wooden sledges, and organized teams of workers have demonstrated that it is physically possible to move large stones without modern machinery, but the specific techniques used at Sacsayhuaman, particularly for the uphill portions of the transport route, remain incompletely understood. Jean-Pierre Protzen, whose careful experimental and archaeological work on Inca stoneworking is the most thorough available, concluded that the basic transport methods were well within the organizational and technical capabilities of the Inca state but that some details of the uphill hauling operations remain unclear. This is the honest state of the evidence: impressive certainty about the organizational framework and the broad engineering approach, genuine uncertainty about some specific operational details. That uncertainty does not justify supernatural explanations; it justifies continued archaeological investigation.

The megalithic monuments of northwestern Europe, the stone circles, passage graves, and standing stone alignments built by the agricultural societies of the Neolithic and early Bronze Age, present a somewhat different profile of questions. Stonehenge, the most famous of these monuments, was constructed in stages over approximately fifteen hundred years, from its earliest earthwork phase around 3000 BCE to its final stone configurations around 1500 BCE, and the engineering of each phase has been studied exhaustively by archaeologists, geologists, engineers, and experimental practitioners. The bluestones of the inner circle, each weighing between two and five tonnes, were transported from the Preseli Hills in Wales, approximately 250 kilometres from the site by the most direct overland route, in the early third millennium BCE. The larger sarsen stones of the outer ring and trilithon horseshoe, weighing between twenty and thirty tonnes each, were brought from Marlborough Downs approximately twenty-five kilometres away. Both transport operations have been studied using experimental methods, and both have been shown to be achievable using timber sledges, ropes, and organized human hauling teams, supplemented possibly by waterborne transport for portions of the bluestone journey.

What the Stonehenge builders lacked in individual stone size they compensated for in the precision of the stone dressing and the geometric sophistication of the layout. The sarsen trilithons were dressed to a consistent cross-section, fitted at their tops with mortise-and-tenon joints that locked the lintels in place, and arranged in a plan that encodes solar and lunar alignments with a precision requiring careful astronomical observation and methodical surveying. The lintels of the outer circle were shaped with a slight curvature to maintain the circular alignment of the ring, a refinement that would have been entirely unnecessary from a structural standpoint and that can only be explained as a deliberate geometric decision. These are not the marks of a culture working at the limits of its capability; they are the marks of a culture that had fully mastered the engineering requirements of its task and was adding refinements beyond strict necessity because the quality of the finished monument mattered to them.

Across all of these traditions, from Baalbek to Sacsayhuaman to Stonehenge, a consistent pattern emerges. The engineering challenges presented by monumental stone construction are real, the organizational demands are formidable, and the precision achieved in the finest examples is genuinely extraordinary. None of it, however, requires explanations beyond the organizational capacity of complex societies, the accumulated craft knowledge of specialist traditions, and the kind of sustained collective investment that only cultures with strong motivating frameworks, religious, political, or social, could maintain across the years and decades that monumental construction demanded. The stones are where they are because people put them there. The wonder of these monuments is not diminished by understanding how they were built; it is deepened, because understanding replaces the cheap astonishment of the inexplicable with the richer admiration of the genuinely understood.

What was lost from the monumental masonry traditions of the ancient world was not the physical knowledge of how to move and shape stone, which is recoverable in principle and has been demonstrated repeatedly by experimental practitioners. What was lost was the organizational and motivational infrastructure that made large-scale stone construction possible as a sustained enterprise. The Inca masonry tradition did not survive the Spanish conquest of the sixteenth century, not because the pirca forgot how to fit stones but because the state system that organized and compensated their labour was destroyed, the religious motivations that drove the building program were suppressed, and the economic resources that fed large construction workforces were redirected. The craft knowledge persisted in the hands of individuals; the conditions for deploying it at scale did not. That distinction is, by now, a familiar one. It is the distinction that runs through the entire history of ancient technical achievement, and it remains as true of the great stone traditions as it is of Damascus steel and Roman concrete.

Eleven

The Libraries That Vanished and the Knowledge That Died

In 415 CE, a woman named Hypatia was murdered in the streets of Alexandria by a mob whose motivations mixed religious faction, political grievance, and the particular fury that crowds direct at those who represent something they want to destroy. Hypatia was a mathematician, an astronomer, and the head of the Neoplatonist philosophical school in Alexandria, the daughter of the mathematician Theon and, by the accounts of contemporaries, among the most learned people in the city that was still, despite centuries of political turbulence, the intellectual center of the Mediterranean world. Her death is often cited in popular histories as a symbolic moment in the destruction of ancient learning, and there is something to the symbolism, though the reality of how ancient knowledge was lost is considerably more complex and considerably more instructive than any single dramatic event could capture.

The Library of Alexandria is the most famous victim in the standard narrative of lost ancient knowledge, and its story has been told so many times, in so many simplified versions, that separating the historical reality from the accumulated mythology has become a task requiring deliberate effort. The library was founded in the early third century BCE under Ptolemy I Soter and developed under his son Ptolemy II Philadelphus into the most ambitious scholarly institution in the ancient world, a repository of texts collected systematically from across the Greek-speaking world and beyond, staffed by scholars engaged in the critical editing, translation, and commentary work that transformed a collection of papyrus rolls into an active intellectual enterprise. The catalog compiled by the scholar Callimachus in the third century BCE listed the library's holdings by subject and author across multiple volumes, and ancient sources give estimates of the total holdings ranging from four hundred thousand to seven hundred thousand rolls, though these figures are difficult to evaluate without knowing how ancient librarians counted rolls that might contain multiple works or portions of works.

The popular image of the library's destruction as a single catastrophic burning at the hands of Julius Caesar, or later of Theophilus, or later still of the Arab general Amr ibn al-As, reflects a desire for narrative simplicity rather than historical accuracy. Caesar did accidentally burn ships in the harbour of Alexandria in 48 BCE, and fire spread to some waterfront buildings including possibly a warehouse of books intended for export, but the scholarly consensus is that this event, whatever its immediate damage, was not the destruction of the main library. The library suffered from neglect, reduced funding, political disruption, and the gradual withdrawal of the patronage networks that had sustained it as the Ptolemaic kingdom was absorbed into the Roman empire and as Alexandria's political centrality declined. It survived in some form into the late Roman period, probably diminished from its Ptolemaic peak, and what finally ended it was not fire but the slow attrition of a changing world in which the specific combination of royal patronage, scholarly community, and institutional investment that had created it no longer existed.

The knowledge housed in the Library of Alexandria was not lost entirely, even in the most pessimistic accounting. Much of what was in the library survived through copying, through the circulation of texts in other institutional collections, through the efforts of individual scholars who preserved texts they valued, and through the translation traditions of the Byzantine, Islamic, and eventually medieval European scholarly worlds. The works of Euclid, Archimedes, Ptolemy, and Apollonius survived in Arabic and Byzantine copies made from texts that ultimately derived from the Alexandrian collection.

The plays of three of the great Athenian tragic poets, Aeschylus, Sophocles, and Euripides, survive because Byzantine scholars made copies that were in turn copied and recopied across the medieval period. What was lost is unquantifiable by definition, because we cannot catalog what has not survived, but the magnitude of the loss can be inferred from the gaps: of the more than one hundred plays attributed to Aeschylus, seven survive; of Sophocles, seven of over one hundred; of the comedies of Menander, who was considered in antiquity to be among the greatest Greek comic writers, enough has been recovered from papyrus fragments and medieval sources to confirm that we have lost the vast majority of his work.

The loss of technical and scientific texts is even harder to assess because the surviving record is so partial that we cannot reliably identify the gaps. We know from references in surviving works that there were ancient texts on metallurgy, on architectural practice, on agricultural chemistry, on military engineering, and on dozens of other practical subjects that have not survived in any form. Vitruvius mentions earlier writers on architecture whose works he drew on and whose names are otherwise unknown to us. Hero of Alexandria describes devices whose full construction details are not always recoverable from the surviving text alone. Pliny the Elder's Natural History quotes and summarizes hundreds of earlier sources on topics ranging from mineralogy to pharmacology, most of whose original works have not survived. The technical literature of the ancient world was extensive, specialized, and, in retrospect, far more fragile than the philosophical and literary texts that commanded the most prestige and attracted the most intensive copying.

The reason for this differential survival is not accidental. The texts that survived best were those that attracted continuous scholarly attention across the centuries of transmission, that were considered culturally or religiously essential, that were valued by the institutional communities, monasteries, cathedral schools, Islamic madrasas, and Byzantine imperial libraries, that maintained the copying tradition. Homer survived because

Greek cultural identity was inseparable from Homer. The Bible survived because the entire institutional apparatus of Christianity was committed to its preservation and reproduction. Euclid survived because his geometry was considered the foundation of mathematical education across the Greek, Islamic, and eventually Western European scholarly traditions. Technical manuals on the preparation of specific dyes, the construction of specific machines, the cultivation of specific crops under specific conditions, had no such institutional advocates. They were copied when someone who needed them found a copy and valued it enough to have it reproduced. When that chain of motivated copying was broken, the text was lost.

The fragility of the physical medium compounded the institutional fragility of the transmission chain. Papyrus, the standard writing material of the classical Mediterranean world, is highly vulnerable to moisture, insects, and the ordinary physical deterioration of organic material over time. A papyrus roll stored in the dry climate of Egypt might survive for millennia, as the papyri recovered from Egyptian sand have demonstrated. A papyrus roll stored in the damp climate of Rome or Athens or Alexandria itself would deteriorate within a few generations without active maintenance and replacement by copying onto fresh material. The transition from papyrus to parchment, which was more durable and better suited to the codex format that gradually replaced the roll from the second century CE onward, helped somewhat, but the bottleneck was not primarily the durability of individual manuscripts but the continuity of the copying tradition that kept them alive as living texts rather than decaying artifacts.

The Islamic translation movement of the eighth through tenth centuries CE, centered on the House of Wisdom in Baghdad and associated with the patronage of the Abbasid caliphate, represents one of the most important acts of intellectual preservation in the history of human knowledge.

Working primarily from Greek originals and from earlier Syriac translations, Arab scholars translated an enormous body of Greek philosophical, scientific, and mathematical texts into Arabic, creating the copies through which much of what survives from the classical world reached the medieval West. Without this movement, the works of Aristotle, which were so central to the intellectual revolution of the twelfth and thirteenth century European universities, might well have been lost entirely. The Arab scholars who carried out this work were not passive transcribers; they engaged critically with the texts they translated, added commentaries, identified contradictions, and in many cases extended the original work in ways that represented genuine intellectual advances. Avicenna's medical synthesis, al-Khwarizmi's algebra, Ibn al-Haytham's optics: these were built on Greek foundations preserved and transmitted through the translation movement, transformed into something new by minds that engaged with them seriously rather than merely reproducing them.

The Byzantine empire played a parallel role in the eastern transmission chain. Byzantine scholars maintained the copying tradition for Greek texts through the medieval period, and the diaspora of Greek scholars from Constantinople to Italy following the Ottoman conquest of 1453 CE brought manuscripts to the West at exactly the moment when the Italian Renaissance was creating the demand for them. The humanists of the fifteenth century, Petrarch, Boccaccio, Poggio Bracciolini, who searched monastery libraries across Europe for forgotten classical texts, were recovering works that had survived at the margins of the Latin tradition, preserved in institutions that had maintained them without fully understanding or valuing what they held. Poggio's discovery of a complete copy of Lucretius's De rerum natura in a German monastery in 1417 CE, and the subsequent influence of that text on Renaissance and early modern thought, is one of the most dramatic individual instances of textual recovery in the Western tradition, a work that had been effectively unknown for centuries suddenly re-entering intellectual circulation with enormous consequences.

Technical knowledge, as distinct from literary, philosophical, and mathematical knowledge, followed different transmission pathways that were in some respects more resilient and in other respects more fragile than the textual tradition. Craft knowledge that lived in the workshop rather than in written texts was vulnerable to the disruption of workshops and craft communities but was also capable of surviving in practice long after the texts that described it had been lost. The glassblowers of Murano, the papermakers of Fabriano, the silk weavers of Lyon, all maintained technical traditions whose origins lay in the ancient or medieval world, transmitted through apprenticeship across generations without systematic textual documentation. When those traditions were eventually studied and documented by historians of technology and materials scientists, researchers found in them layers of accumulated empirical knowledge that no surviving text had recorded.

The scribes who copied manuscripts in the scriptorium of a medieval monastery were engaged in a form of knowledge preservation that had its own vulnerabilities and its own characteristic losses. A scribe copying a text in a language he did not fully understand, from a damaged original with uncertain readings, in conditions of poor light and inadequate heating, produced a copy that preserved the original imperfectly and introduced new errors that subsequent copyists would in turn copy and extend. The transmission of technical texts was particularly vulnerable to this process because a small misreading of a number, a proportion, or a procedural step could render an instruction useless or actively misleading, and a copyist with no practical knowledge of the subject could not detect the error. The history of ancient recipes and technical instructions transmitted through the medieval copying tradition is full of passages that are garbled beyond reconstruction, where a specific and probably workable ancient technique has been reduced to an incomprehensible sequence of operations by the cumulative errors of scribes who did not know what they were copying.

The destruction of knowledge was not always passive or accidental. Libraries and archives were deliberately burned as acts of conquest and cultural suppression throughout the ancient and medieval world. The burning of the Persian royal library at Persepolis by Alexander the Great in 330 BCE, whatever its motivation, represented the destruction of a significant collection of administrative, religious, and possibly scientific texts from the Achaemenid tradition. The Qin emperor Shi Huangdi's notorious burning of books in 213 BCE targeted texts of philosophy, history, and poetry considered politically threatening to the new imperial ideology, and while the tradition exaggerates the completeness of the destruction, there is no doubt that it caused real and permanent losses. The systematic destruction of Aztec and Maya codices by Spanish missionaries in the sixteenth century, motivated by the conviction that indigenous texts were instruments of devil worship, eliminated most of the written record of Mesoamerican astronomical, mathematical, and historical knowledge, leaving only four Maya codices and a handful of Aztec manuscripts out of what had been an extensive literary and scientific tradition.

Diego de Landa, the Franciscan friar who oversaw the burning of the Maya books at Mani in 1562, later expressed regret at the scale of the destruction while maintaining that it had been necessary for the spiritual welfare of the people whose books he had burned. He subsequently wrote an account of Maya culture and language, the Relacion de las Cosas de Yucatan, that has become one of the most important sources for understanding the civilization whose records he helped destroy. The bitter irony of this situation, the destroyer becoming a primary source for knowledge of what was destroyed, captures something essential about the history of lost knowledge: the people who caused the losses were not always unaware of the value of what they were destroying, and the knowledge that survived often did so through the very channels that the destruction had opened.

What this survey of lost knowledge makes clear, finally, is that the loss was not the result of a single catastrophe or a single failure of civilization. It was the accumulated result of thousands of individual losses, each one contingent on specific circumstances of political disruption, institutional collapse, physical deterioration, deliberate destruction, or simple failure of motivated transmission. Some of those losses were irreversible. The technical knowledge of the best Damascus steel smiths, the full construction details of Hero's automata, the chemical formula of Greek Fire, the complete astronomical tradition of the Antikythera Mechanism's makers: these belong to the category of losses that no amount of archaeological or scientific ingenuity will fully reverse, because the knowledge lived in practices and people rather than in texts, and the practices and the people are gone.

Other losses have been or may yet be recovered, through the patient work of archaeologists who find new texts in desert sands or monastery archives, through materials scientists who read the chemistry of ancient objects and reconstruct what their makers knew, through experimental practitioners who rebuild what texts and objects describe and discover in the process what the descriptions left out. The boundary between recoverable and irrecoverable loss shifts with every new discovery, and the history of the past two centuries gives genuine grounds for optimism about how much more may yet be found.

Twelve
What We Still Do Not Know

In 2006, a team of archaeologists excavating a site in the Bavarian city of Tuttlingen found a fragment of bone flute in a layer dated to approximately forty-three thousand years before the present. The instrument was carved from a vulture's wing bone, with five finger holes cut with a precision suggesting a maker who understood how hole placement relates to the pitch of the notes produced. It is among the oldest known musical instruments in the world, and its existence raises a question that runs through all of archaeology: if people were making tuned musical instruments forty-three thousand years ago, what else were they doing that has not survived the passage of time? Bone and fired clay endure; wood and fibre and leather do not. The record we have of human technical achievement before the invention of writing is almost certainly a severe underrepresentation of what was actually accomplished, because the materials that survive to be excavated are a biased sample of the full range of what ancient people made.

This is the fundamental epistemic problem facing anyone who attempts to assess what the ancient world knew and did not know. Absence of evidence is not evidence of absence, a principle that the history of discovery repeatedly confirms. The Antikythera Mechanism was not expected to exist until it was found. The sophistication of Inca polygonal masonry was not fully appreciated until researchers began measuring its tolerances with modern instruments. The self-healing chemistry of Roman marine concrete was not understood until twenty-first-century materials scientists looked at two-thousand-year-old harbour structures with synchrotron X-ray facilities. In each case, the discovery did not reveal that the ancient world had been doing something impossible; it revealed that the modern world had been underestimating what the ancient world was capable of. The pattern is consistent enough to warrant a general principle: assume less certainty about what ancient civilizations could not do, and maintain a greater openness to the possibility that the surviving record is an incomplete guide to the full extent of their capabilities.

The ongoing revolution in ancient DNA analysis is producing findings that require constant revision of the population history of human prehistory, and with that population history comes the implicit revision of the history of technical knowledge transmission. When ancient DNA analysis revealed that Neanderthals and anatomically modern humans interbred at multiple points over tens of thousands of years, it became necessary to reconsider whether technical knowledge might have been exchanged between these populations in ways that the existing archaeological record did not clearly indicate. When genomic data revealed unexpected migrations and admixtures in Bronze Age Europe and the Americas, it opened questions about whether particular technical traditions had spread with the migrating populations or had developed independently. The history of ancient technology is entangled with the history of ancient human movement in ways that the physical record of tools and structures can only partially illuminate.

The deep-sea archaeology program that recovers ancient shipwrecks is one of the most productive frontiers of discovery for the history of ancient technology, for reasons that the Antikythera Mechanism illustrates perfectly. The sea preserves in anaerobic, cold, dark conditions objects that would have decayed within decades on land. An ancient cargo ship carrying bronze instruments, glass vessels, or complex mechanical objects is a sealed time capsule, and the number of such wrecks recoverable from the Mediterranean, the Black Sea, the Red Sea, and the waters of the Indian Ocean trade routes is very large. Current estimates suggest that tens of thousands of ancient shipwrecks remain on the seafloor of the Mediterranean alone, the great majority of them unexcavated. What they might contain, what instruments, machines, artworks, or texts preserved in sealed containers against the two-thousand-year passage of time, is genuinely unknown. The Antikythera wreck yielded its mechanism almost by accident; a systematic program of deep-sea excavation using the remotely operated vehicle technology now available might yield discoveries whose implications for understanding ancient technology are equally transformative.

The question of ancient astronomical knowledge is one of the areas where the gap between what is known and what may remain to be discovered is most conspicuous. The Antikythera Mechanism demonstrated that by the second or first century BCE, Greek astronomers had translated their mathematical models of celestial motion into mechanical form with extraordinary precision. The astronomical knowledge embedded in the device, including the Metonic cycle, the Saros cycle, and the Hipparchan model of lunar motion, was derived from observations accumulated over centuries, some of it originating in the Babylonian astronomical tradition that predated the Greek by more than a thousand years. The full extent of Babylonian astronomical knowledge, which is preserved in cuneiform tablets that have been only partially studied and translated, continues to yield surprises. Researchers working on the Babylonian astronomical corpus have found evidence that Babylonian astronomers used arithmetical procedures equivalent to early forms of calculus to track the motion of Jupiter as early as the fourth century BCE, a mathematical sophistication that was not previously attributed to ancient Mesopotamian astronomy. The tablets that contain this evidence had been sitting in museum collections since the nineteenth century, examined but not fully understood.

The history of ancient medicine is another domain where the surviving record almost certainly underrepresents both the extent and the sophistication of ancient knowledge. The pharmacological traditions of Greece, Rome, India, and China documented in surviving texts describe the medical use of hundreds of plant, animal, and mineral substances, many of which have subsequently been found to contain compounds with genuine pharmacological activity. The willow bark that ancient healers prescribed for pain and fever contains salicylic acid, the active principle of aspirin. The foxglove preparations used in traditional European medicine for heart conditions contain digitalis glycosides, now used in modern cardiology. The antimalarial properties of artemisia plants, used in Chinese traditional medicine for more than two thousand years, were confirmed by modern pharmacological research in the 1970s and led to the development of artemisinin-based therapies that are now the primary treatment for drug-resistant malaria. These are not coincidences. They are the residue of long empirical traditions in which ancient practitioners systematically observed and recorded the effects of natural substances on human disease, accumulating a pharmacological knowledge whose mechanisms they could not explain but whose efficacy they had thoroughly tested across generations of practice.

The question of what we are still missing from ancient pharmacological knowledge is more than academic. The biodiversity crisis of the present era is eliminating plant and animal species at a rate unprecedented in human history, and with those species goes any possibility of studying the substances that ancient healers used and that modern pharmacology has not yet had the opportunity to investigate. Some of the plant species mentioned in ancient medical texts as sources of important remedies are no longer clearly identifiable, either because they have become extinct, because their ancient names no longer correspond to any currently recognized species, or because the populations of the relevant species have changed genetically over two millennia of selection pressure. The race between archaeobotanical reconstruction and species extinction is one of the less-discussed but genuinely consequential practical stakes of the project of recovering ancient technical knowledge.

Ancient materials science is another area of active discovery and genuine remaining uncertainty. The research into Roman concrete has demonstrated that ancient construction materials sometimes performed in ways that modern materials science is only beginning to understand. Similar investigations are underway into other ancient materials whose properties are not fully explained by conventional analysis. The purple dye of the ancient world, Tyrian purple, derived from the mucus glands of two species of predatory marine snails of the genus Murex, was the most valuable colorant in the ancient Mediterranean economy, worth more than its weight in gold by weight of the finished fabric, and its production supported an extensive industry across the Phoenician coast and later across the Roman world.

The chemistry of Tyrian purple, 6,6-dibromoindigo, is now understood, but the full production process, from snail harvest through fermentation to dye bath to finished fabric, incorporated empirical refinements that ancient dyers had accumulated over centuries and that modern reconstructions have found difficult to reproduce at the quality level of the finest ancient examples. Whether there are aspects of the ancient production process, specific fermentation conditions, specific mineral additives, specific physical manipulations of the dye bath, that have not been fully recovered, remains an open question in the archaeology of ancient technology.

The ancient world's knowledge of materials we would now classify as adhesives, lubricants, preservatives, and surface treatments represents another domain of likely significant loss. Ancient boat builders used pine pitch, beeswax, various animal fats, and mineral compounds to seal and protect wooden hulls in ways whose specific formulations varied by region and tradition. Some of these formulations, including the pitch compounds used on Egyptian and Phoenician vessels, have been analyzed from surviving residues and appear to have had properties, particularly long-term resistance to marine boring organisms, that modern wood preservation treatments do not fully replicate. The Roman opus signinum, a lime-based waterproof cement incorporating crushed terracotta that was used to line cisterns, baths, and water channels, has been studied and found to incorporate chemical refinements that give it a durability in wet environments beyond what simple lime and clay would produce. Whether these refinements were deliberate or the result of empirical optimization through long trial and error, and what specific aspects of the ancient production process produced them, are questions that continue to generate research.

What all of these open questions share is that they involve knowledge that was once possessed, was either fully or partially lost, and has the potential to be recovered through the application of modern scientific and analytical methods to ancient materials and texts. The recovery is not guaranteed; some of what was lost is beyond the reach of any analytical technique because it lived in practices and perceptions that left no material trace. But the track record of the past century of archaeological and materials science research gives genuine reason to believe that significant discoveries remain to be made, and that the picture of ancient technical achievement that will be available to researchers a century from now will be considerably richer and more surprising than the one available today. Every generation of archaeologists has found things that overturned the assumptions of the previous generation, and there is no good reason to believe that this process is approaching its end.

The honest acknowledgment of what we do not know is not a counsel of despair but a reminder of the appropriate posture before the past. The ancient world was more capable than we have assumed. The losses it suffered were more extensive than we can fully measure. The recoveries already achieved are more remarkable than they might seem to readers who have not followed the specific stories of individual discoveries, and the recoveries still possible are, on the evidence of the past century, likely to be remarkable in ways that are genuinely difficult to predict. The past is not finished. It is still being excavated, analyzed, and understood, and the process yields, repeatedly and without sign of slowing, the particular kind of surprise that comes from discovering that human beings, in circumstances very different from our own, found ways to do things we had not imagined they could do.

Conclusion

Why Ancient Ingenuity Still Matters

The twelve technologies and technical traditions examined in this book are not a complete inventory of ancient achievement. They are a set of windows, chosen because each one illuminates something distinct about the relationship between human ingenuity, material knowledge, and historical memory. What they reveal, collectively, is a past considerably richer and more technically sophisticated than the version of ancient history that most people carry in their heads. The Greek astronomers who built the Antikythera Mechanism were not primitive philosophers fumbling toward modern science; they were professional mathematicians and engineers working within a tradition of accumulated astronomical and mechanical knowledge that produced results we are still working to fully understand. The Roman builders who poured volcanic ash into seawater to create a self-healing concrete were not lucky amateurs stumbling upon a fortunate mixture; they were systematic empiricists who had spent centuries learning, through observation and practice, what their specific materials could do under specific conditions.

The Damascus smiths who stood at their forges for hours watching the colour of wootz steel shift from cherry red toward straw yellow, feeling in the sound and resistance of the metal under their hammers the precise moment when the carbide network was forming correctly, were not primitive craftsmen working by superstition; they were highly trained professionals applying a body of embodied technical knowledge so sophisticated that modern metallurgists using electron microscopes have taken decades to understand what the smiths were actually producing. The Byzantine engineers who calibrated their Greek Fire siphon systems to project sustained streams of burning naphtha at enemy warships were not lucky inheritors of a mysterious formula; they were the product of a military-industrial tradition that had developed, tested, and refined both the chemical formulation and the delivery system over generations of active use. And the Nasca surveyors who stretched cords across a Peruvian desert plateau to establish the geometry of a condor figure several hundred metres long were not the agents of alien visitors; they were skilled practitioners of a planar geometry tradition applied with extraordinary patience and discipline to the specific demands of their ritual landscape.

The pattern across all twelve is the same, and it is worth stating plainly because popular culture consistently obscures it. Ancient technical achievement was not accidental, not supernatural, not the product of a lost civilization more advanced than our own that seeded its knowledge across the world. It was the product of human intelligence applied patiently and systematically within specific material and cultural contexts, accumulating knowledge through empirical observation and practical experiment across generations of practitioners who built on each other's work. It was, in other words, recognizably the same thing as modern technical achievement, operating with different tools, different materials, different theoretical frameworks, and different cultural motivations, but driven by the same fundamental human capacity for observation, inference, experiment, and the transmission of learned skill from one generation to the next.

The loss of so much of this knowledge is the second great theme of this book, and it demands a different emotional register: not admiration but unease, not wonder but something closer to grief. The muqannis who understood the hydrogeology of the Iranian piedmont better than any university-trained hydrologist of the early modern era, the Damascus smiths who were producing carbon nanotube-reinforced steel without knowing it, the Nasca surveyors who could project a geometrically coherent figure across several hundred metres of uneven desert floor using nothing but cord and stakes and accumulated skill, the anonymous engineers who built the Antikythera Mechanism and the dozens or hundreds of similar devices that must have existed and have not survived: all of them carried knowledge that was irreplaceable, and all of them belonged to traditions that the ordinary processes of history dismantled without any particular malice or awareness of what was being lost.

The mechanisms of that loss are thoroughly ordinary. Trade routes were disrupted and the raw materials that specific craft traditions depended on became unavailable. Political systems that had organized and funded large-scale technical enterprises collapsed and were replaced by systems with different priorities and different resource allocations. Cities that had been centers of technical learning were destroyed, abandoned, or transformed in ways that ended the workshop and educational traditions that had made them productive. Languages in which technical knowledge was recorded died as living languages, becoming inaccessible without specialized training that relatively few people possessed. Craft masters died without training successors because the economic conditions that had made their skills valuable had changed. Libraries and archives were burned, damaged, or simply neglected until their contents deteriorated beyond recovery. None of these processes is exotic or mysterious. All of them are the normal conditions of historical change, operating on the specific vulnerabilities of technical knowledge traditions that required continuous active support to survive.

Understanding this is important not as historical trivia but as a practical lesson about the nature of knowledge itself. The modern world is not immune to the processes that erased ancient technical traditions. The specific vulnerabilities are different, but the fundamental dependence of knowledge on the social and material conditions that support it is the same. When a specialist craft tradition is economically unviable and practitioners retire without training successors, the knowledge they carry dies with them regardless of whether it has been documented. When a critical infrastructure system becomes dependent on a small number of highly specialized technicians whose training takes decades and who cannot be quickly replaced, the knowledge required to maintain that system is vulnerable to exactly the kind of disruption that ended ancient craft traditions. When digital archives store data in formats that become obsolete faster than the institutions maintaining them can keep pace, the result is a modern version of the ancient problem of texts that survive physically but become unreadable for lack of the interpretive tradition that gave them meaning.

None of this is an argument for pessimism about the modern world's ability to preserve its knowledge. It is an argument for taking the preservation question seriously rather than assuming that the complexity and redundancy of modern information systems provides automatic protection against the kinds of loss that the ancient world experienced. The Roman engineers who built the aqueducts did not expect their political and economic system to collapse in ways that made maintenance impossible. The Byzantine authorities who guarded the Greek Fire formula did not expect the state system that gave the formula its meaning and application to be destroyed. The Alexandrian scholars who worked in the greatest library the ancient world had ever assembled did not plan for the gradual erosion of the institutional conditions that made their work possible. The assumption that the conditions that currently support a technical tradition will continue to exist is precisely the assumption that historical experience most consistently refutes.

The third theme that runs through this book, alongside admiration for ancient achievement and unease at the fragility of knowledge, is the continuing productivity of the encounter between the modern world and the ancient one. The story of ancient technology is not a closed story. The Antikythera Mechanism is still yielding new information to imaging technologies that did not exist when it was first discovered. Roman marine concrete is generating active engineering research into whether its principles can be adapted for modern sustainable construction. The Baghdad Battery is still the subject of serious scholarly debate whose resolution may reveal something important about the epistemological standards we apply to ambiguous archaeological evidence. The Nazca lines are being remapped by satellite imaging that reveals new figures previously unknown. Ancient pharmacological traditions are being systematically screened for compounds with potential medical applications. Egyptian stoneworking techniques are being reconstructed through experimental archaeology that continues to refine our understanding of what ancient craftsmen could achieve with copper tools and organized labour.

Each of these ongoing encounters between modern knowledge and ancient achievement is productive in both directions. Modern science illuminates what ancient practitioners were doing and why it worked. Ancient achievement challenges modern assumptions about the range of what human ingenuity can accomplish without the theoretical frameworks and industrial infrastructure that the modern world considers necessary. The encounter is not one of superiority and inferiority in either direction; it is a genuine dialogue between different forms of knowledge, each with capabilities the other lacks, and the dialogue is richer and more surprising the more carefully it is conducted.

There is also a specifically contemporary relevance to the question of how ancient civilizations lost technical knowledge that this book has only gestured toward but that deserves explicit acknowledgment. The climate and ecological disruptions of the present era are introducing forms of systemic stress into the social and material systems that support modern technical traditions, and the history of ancient knowledge loss provides some indication of what kinds of stress are most damaging to the continuity of complex technical knowledge. Disruptions that interrupt supply chains for critical materials, that destroy the economic conditions that make specialist craft traditions viable, that undermine the institutional frameworks that support technical education and research, that reduce the redundancy of knowledge systems by concentrating expertise in fewer hands and fewer places: these are exactly the conditions that ended ancient technical traditions, and they are conditions that various forms of systemic disruption can produce in modern contexts as readily as in ancient ones.

This is the deepest reason why ancient ingenuity still matters. Not as an object of nostalgic admiration, not as evidence for the implausible claim that ancient people were more capable than modern ones, but as a record of what human intelligence can accomplish across a very wide range of material and institutional contexts, and as a cautionary account of how even the most impressive technical achievements can be lost when the conditions that sustained them change faster than the knowledge traditions can adapt. The people who made the objects and built the structures discussed in this book were not fundamentally different from the people reading about them now. They were curious, skilled, ambitious, competitive, and proud of what they could make. They made things that astonish us. They also lost things that we can only partly recover, and they lost them through processes that remain entirely operative in the world today.

The appropriate response to this history is neither complacency nor despair but attention. Attention to what is being maintained and what is being allowed to erode. Attention to which knowledge traditions are fragile and what conditions support their survival. Attention to the gap between what exists in texts or databases and what exists in the hands and minds of people who actually know how to do things. Attention, above all, to the principle that the ancient world demonstrates with uncomfortable consistency: that knowledge is not self-preserving, that achievement is not self-sustaining, and that the civilization which assumes its technical inheritance is secure without active effort to maintain it is making exactly the mistake that the best ancient civilizations made, and that history has not yet finished punishing.

The gears of the Antikythera Mechanism are still in Athens, still being studied, still capable of surprise. The Roman harbours are still in the Mediterranean, still hardening in the sea. The Nazca lines are still in the desert, still accumulating the meaning that nine centuries of makers gave them. And the questions they raise, about what human beings can accomplish, about how much has been lost, and about what it takes to keep knowledge alive across the generations, are still, in every sense that matters, entirely open.

Select Bibliography

Aristotle. Meteorologica. Translated by H. D. P. Lee. Loeb Classical Library. Harvard University Press, 1952.

Cobb, Paul M. The Race for Paradise: An Islamic History of the Crusades. Oxford University Press, 2014.

Craddock, Paul T. Early Metal Mining and Production. Edinburgh University Press, 1995.

Fitchen, John. Building Construction Before Mechanization. MIT Press, 1986.

Freestone, Ian, et al. "The Portland Vase: The Origin of the Blank." Archaeometry 32, no. 2 (1990): 111–128.

Haldon, John. "Greek Fire Revisited: Recent and Current Research." In Byzantine Style, Religion and Civilization, edited by Elizabeth Jeffreys. Cambridge University Press, 2006.

Hodges, Henry. Technology in the Ancient World. Alfred A. Knopf, 1970.

Jackson, Marie D., et al. "Accretionary Mineral Deposits in Roman Marine Concrete." American Mineralogist 102 (2017): 1435–1450.

Komnene, Anna. The Alexiad. Translated by E. R. A. Sewter. Penguin Classics, 2009.

Lehner, Mark. The Complete Pyramids. Thames and Hudson, 1997.

Mayor, Adrienne. Greek Fire, Poison Arrows and Scorpion Bombs: Biological and Chemical Warfare in the Ancient World. Overlook Duckworth, 2003.

Needham, Joseph. Science and Civilisation in China. Volume 5: Chemistry and Chemical Technology. Cambridge University Press, 1974.

Paufler, Peter, et al. "Damascus Swords with Carbon Nanotubes." Nature 444 (2006): 286.

Price, Derek de Solla. Gears from the Greeks: The Antikythera Mechanism, a Calendar Computer from ca. 80 BC. Science History Publications, 1974.

Roland, Alex. "Greek Fire: A Re-Evaluation." Journal of Military History 58, no. 1 (1994): 1–18.

Stocks, Denys A. Experiments in Egyptian Archaeology: Stoneworking Technology in Ancient Egypt. Routledge, 2003.

Theophanes. The Chronicle of Theophanes Confessor. Translated by

Cyril Mango and Roger Scott. Clarendon Press, 1997.

Vitruvius. On Architecture (De Architectura). Translated by Frank Granger. Loeb Classical Library. Harvard University Press, 1931.

Wulff, Hans E. The Traditional Crafts of Persia. MIT Press, 1966.

About the Author

Kayla Brooks is a writer and researcher based in the United Kingdom with a long-standing interest in the history of science, ancient civilizations, and the ways in which technical knowledge is made, transmitted, and lost. She has spent many years reading across the fields of archaeology, materials science, engineering history, and classical studies, and she writes for readers who share her conviction that the ancient world deserves to be understood on its own terms rather than filtered through the assumptions of the modern one.

The Lost Technologies of the Ancient World is her first book.

www.ingramcontent.com/pod-product-compliance
Lightning Source LLC
LaVergne TN
LVHW100306210926
848480LV00003B/1036